THE ACCIDENTAL MIND

THE ACCIDENTAL MIND

DAVID J. LINDEN

The Belknap Press of Harvard University Press

Cambridge, Massachusetts · London, England

2007

Printed in the United States of America

Library of Congress Cataloging-in-Publication Data

Linden, David J., 1961–
The accidental mind / David J. Linden.
p.   cm.
Includes bibliographical references and index.
ISBN-13: 978-0-674-02478-6 (cloth : alk. paper)
ISBN-10: 0-674-02478-8 (cloth : alk. paper)
1. Brain—Popular works.   2. Brain—Evolution.
3. Neuropsychology.   I. Title.

QP376.L577 2007
612.8′2—dc22   2006047905

For Herbert Linden, M.D.

Contents

THE ACCIDENTAL MIND

The large brain, like large government, may not be able to do simple things in a simple way.

—Donald O. Hebb

Now, the president says that the jury is out on evolution . . . Here in New Jersey, we're countin' on it.

—Bruce Springsteen

Brain, Explained

THE BEST THING about being a brain researcher is that, in a very small number of situations, you can appear to have the power of mind reading. Take cocktail parties. Chardonnay in hand, your host makes one of those introductions where he feels compelled to state your occupation: "This is David. He's a brain researcher." Many people are wise enough to simply turn around at this point and go looking for the bourbon and ice. Of those who stay behind about half can be counted on to pause, look heavenward, and raise their eyebrows in preparation for speech. "You're about to ask if it's true that we only use 10 percent of our brain, aren't you?" Wide-eyed nodding. An amazing episode of "mind reading."

Once we get past the 10-percent-of-the-brain thing (which, I should mention, has no basis in reality), it becomes clear that many people have a deep curiosity about brain function. Really fundamental and difficult questions come up right away:

"Will playing classical music to my newborn really help his brain grow?"

"Is there a biological reason why the events in my dreams are so bizarre?"

"Are the brains of gay people physically different from the brains of straight people?"

"Why can't I tickle myself?"

These are all great questions. For some of them, the best scientific answer is fairly clear and for others it is somewhat evasive (me, in my best Bill Clinton voice: "What exactly do you mean by "brain"?). It's fun to talk to non–brain researchers about these kinds of things because they are not afraid to ask the hard questions and to put you on the spot.

Often, when the conversation is over, people will ask, "Is there a good book on brain and behavior for a nonspecialist audience that you can recommend?" Here, it gets tricky. There are some books, such as Joe Le Doux's *Synaptic Self,* that do a great job on the science, but that are rough sledding unless you've already got a college degree in biology or psychology. There are others, such as Oliver Sacks's *Man Who Mistook His Wife for a Hat* and V. S. Ramachandran and Sandra Blakeslee's *Phantoms in the Brain* that tell fascinating and illuminating stories based on case histories in neurology, but that really don't convey a broad understanding of brain function and that largely ignore molecules and cells. There are books that talk about molecules and cells in the brain, but many of them are so deadly dull that you can start to feel your soul depart your body before you finish the very first page.

What's more, many books about the brain, and even more shows on educational television, perpetuate a fundamental misunderstanding about neural function. They present the brain as a beautifully engineered, optimized device, the absolute pinnacle of design. You've probably seen it before: a human brain lit dramatically from the side, with the camera circling it as if taking a helicopter shot of Stonehenge and a modulated baritone voice exalting the brain's elegant design in reverent tones.

This is pure nonsense. The brain is not elegantly designed by any means: it is a cobbled-together mess, which, amazingly, and in spite of its shortcomings, manages to perform a number of very impressive functions. But while its overall function is impressive, its design is not. More important, the quirky, inefficient, and bizarre plan of the brain and its constituent parts is fundamental to our human experience. The particular texture of our feelings, perceptions, and actions is derived, in large part, from the fact that the brain is not an optimized, generic problem-solving machine, but rather a weird agglomeration of ad hoc solutions that have accumulated throughout millions of years of evolutionary history.

So, here's what I'll try to do. I will be your guide to this strange and often illogical world of neural function, with the particular charge of pointing out the most unusual and counterintuitive aspects of brain and neural design and explaining how they mold our lives. In particular, I will try to convince you that the constraints of quirky, evolved brain design have ultimately led to many transcendent and unique human characteristics: our long childhoods, our extensive memory capacity (which is the substrate upon which our individuality is created by experience), our search for long-term love relationships, our need to create compelling narrative and, ultimately, the universal cultural impulse to create religious explanations.

Along the way, I will briefly review the biology background you will need to understand the things I am guessing you most want to know about the brain and behavior. You know, the good stuff: emotion, illusion, memory, dreams, love and sex, and, of course, freaky twin stories. Then, I'll try my best to answer the big questions and to be honest when answers are not at hand or are incomplete. If I don't answer all of your questions, try visiting the book's website, accidentalmind.org. I'll strive to make it fun, but I'm not going to "take all the science out." It will not be, as you might find on a label at Whole Foods, "100 percent molecule free."

Max Delbrück, a pioneer of molecular genetics, said, "Imagine that your audience has zero knowledge but infinite intelligence." That sounds just about right to me, so that's what I'll do. Let's roll.

Chapter One

The Inelegant Design of the Brain

WHEN I WAS IN middle school, in California in the 1970s, a popular joke involved asking someone, "Want to lose 6 pounds of ugly fat?" If the reply was positive it would be met with "Then chop off your head! Hahahaha!" Clearly, the brain did not hold a revered place in the collective imagination of my classmates. Like many, I was relieved when middle school drew to a close. Many years later, however, I have been similarly distressed by the opposite view. Particularly when reading books or magazines or watching educational television, I have been taken aback by a form of brain worship. Discussion of the brain is most often delivered in a breathless, awestruck voice. In these works the brain is "an amazingly efficient 3 pounds of tissue, more powerful than the largest supercomputer," or "the seat of the mind, the pinnacle of biological design." What I find problematic about these statements is not the deep appreciation that mental function resides in the brain, which is indeed amazing. Rather, it is the assumption that since the mind is in the brain, and the mind is a great achievement, the design and function of the brain must then be elegant and efficient. In short, it is imagined by many that the brain is well engineered.

Nothing could be further from the truth. The brain is, to use one of my favorite words, a kludge (pronounced "klooj"), a design that is inefficient, inelegant, and unfathomable, but that nevertheless works. More evocatively, in the words of the military historian Jackson Granholm, a kludge is "an ill-assorted collection of poorly matching parts, forming a distressing whole."

What I hope to show here is that at every level of brain organization, from regions and circuits to cells and molecules, the brain is an inelegant and inefficient agglomeration of stuff, which nonetheless works surprisingly well. The brain is not the ultimate general-purpose supercomputer. It was not designed at once, by a genius, on a blank piece of paper. Rather, it is a very peculiar edifice that reflects millions of years of evolutionary history. In many cases, the brain has adopted solutions to particular problems in the distant past that have persisted over time and have been recycled for other uses or have severely constrained the possibilities for further change. In the words of the pioneering molecular biologist François Jacob, "Evolution is a tinkerer, not an engineer."

What's important about this point as applied to the brain is not merely that it challenges the notion of optimized design. Rather, appreciation of the quirky engineering of the brain can provide insights into some of the deepest and most particularly *human* aspects of experience, both in day-to-day behavior and in cases of injury and disease.

SO, WITH THESE issues in mind, let's have a look at the brain and see what we can discern about its design. What are the organizational principles that emerge? For this purpose, imagine that we have a freshly dissected adult human brain before us now (Figure 1.1). What you would see is a slightly oblong, grayish-pink object weighing about 3 pounds. Its outer surface, which is called the cortex, is covered with thick wrinkles that form deep grooves. The pattern of these grooves and wrinkles looks like it might be variable, like a fingerprint, but it is actually very similar in all human brains. Hanging off the back of the brain is a structure the size of a squashed baseball with small crosswise grooves. This is called the cerebellum, which means "little brain." Sticking out of the bottom of the brain, somewhat toward the back end is a thick stalk called the brainstem. We've lopped off the very bottom of the brainstem where it would otherwise taper to form the top of the spinal cord. Careful observation would reveal the nerves, called the cranial nerves, which carry information from the eyes, ears, nose, tongue, and face into the brainstem.

One obvious characteristic of the brain is its symmetry: the view from the top shows a long groove from front to back that divides the cortex (which means "rind"), the thick outer covering of the brain, into two equal halves. If we slice completely through the brain, using this front-to-back groove as a guide, and then turn the cut side of the right half toward us, we see the view shown in the bottom of Figure 1.1.

Looking at this image makes it clear that the brain is not just a homogeneous blob of stuff. There are variations in shape, color, and texture of the brain tissue across brain regions, but these do not tell us about the functions of these various regions. One of the most useful ways to investigate the function of these locations is to look at people who have sustained damage to various parts of the brain. Such investigations have been complemented by animal experiments in which small regions of the brain are precisely damaged through surgery or the administration of drugs, after which the animal's body functions and behavior are carefully observed.

The brainstem contains centers that control extremely basic regulation of

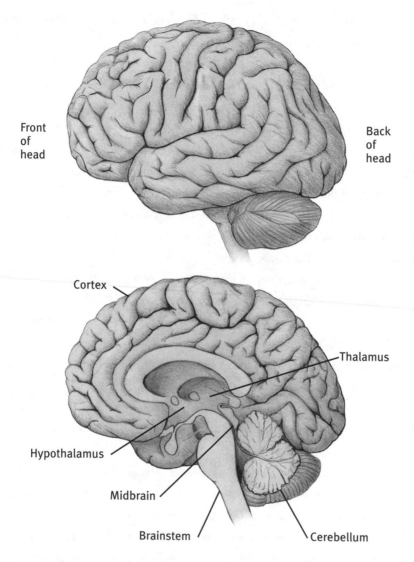

Front
of
head

Back
of
head

Cortex

Thalamus

Hypothalamus

Midbrain

Brainstem

Cerebellum

FIGURE 1.1. The human brain. The top shows the intact brain viewed from the left side. The bottom shows the brain sliced down the middle and then opened to allow the right side to face us. *Joan M. K. Tycko, illustrator.*

the body that are not under your conscious control, including vital functions such as regulation of heart rate, blood pressure, breathing rhythm, body temperature, and digestion. It also contains the control centers for some important reflexes, such as sneezing, coughing, and vomiting. The brainstem houses relays for sensations coming up the spinal cord from your skin and muscles as well as for command signals coming from your brain and destined for muscles in your body. It also contains locations involved in producing feelings of wakefulness versus sleepiness. Drugs that modify your state of wakefulness, such as sleeping pills or general anesthetics on the one hand and caffeine or amphetamines on the other, act on these brainstem regions. If you get a small area of damage in your brainstem (from an injury, tumor, or stroke), you could be rendered comatose, unable to be aroused by any sensation, but extensive damage in the brainstem is almost always fatal.

The cerebellum, which is richly interconnected with the brainstem, is involved with coordination of movements. In particular, it uses feedback from your senses about how your body is moving through space in order to issue fine corrections to the muscles to render your movements smooth, fluid, and well coordinated. This cerebellar fine-tuning operates not only in the most demanding forms of coordination such as hitting a baseball or playing the violin, but also in everyday activities. Damage to the cerebellum is subtle. It will not paralyze you, but rather will typically result in clumsiness in performing simple tasks that we take for granted, such as reaching smoothly to grasp a coffee cup or walking with a normal gait; this phenomenon is called ataxia.

The cerebellum is also important in distinguishing sensations that are "expected" from those that are not. In general, when you initiate a movement and you have sensations which result from that movement, you tend to pay less attention to those sensations. For example, when you walk down the street and your clothes rub against your body, these are sensations that you mostly ignore.

By contrast, if you were standing still and you started to feel similar rubbing sensations on your body, you would probably pay a lot of attention. You would probably whirl around to see who was groping you. In many situations, it is useful to ignore sensations produced by your own motion and pay close attention to other sensations that originate from the outside world. The cerebellum receives signals from those brain regions that create the commands that trigger body motion. The cerebellum uses these signals to predict the sensations that are likely to result from this motion. Then the cerebellum sends inhibitory signals to other brain regions to subtract the "expected" sensations from the "total" sensations and thereby change the way they feel to you.

This may all sound a bit abstract, so let's consider an example. It is well known that you can't tickle yourself. This is not just true in certain cultures; it is worldwide. What's different about having someone else tickle you, which can result in a very strong sensation, and self-tickling, which is ineffective? When researchers in Daniel Wolpert's group at University College, London, placed people's heads in a machine that can make images showing the location and strength of brain activity (called functional magnetic-resonance images, or fMRI) and then tickled them, they found strong activation in a brain region involved in touch sensation called the somatosensory cortex and no significant activation in the cerebellum. When people were then asked to tickle themselves on that same part of the body, it was seen that there was a spot of activation in the cerebellum and reduced activity in the somatosensory cortex. The interpretation of this experiment is that commands to activate the hand motions in self-tickling stimulated the cerebellum, which then formed a prediction of the expected sensation and sent signals encoding this prediction to inhibit the somatosensory cortex. The reduced activation of the somatosensory cortex was then below the threshold necessary to have the sensation feel like tickling. Interestingly, there are now reports that some humans who sustain damage to the

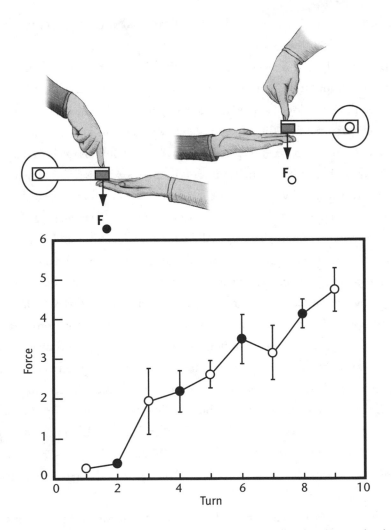

FIGURE 1.2. Force escalation in a tit-for-tat finger-tapping task. The white circles show the force of finger taps delivered by one subject, the black circles the force from the other subject. In 9 tit-for-tat exchanges, the force increased almost 20-fold. Adapted from S. S. Shergill, P. M. Bays, C. D. Frith, and D. M. Wolpert, Two eyes for an eye: the neuroscience of force escalation, *Science* 301:187 (2003); copyright 2003 AAAS. *Joan M. K. Tycko, illustrator.*

cerebellum cannot generate predictions of expected sensations and therefore can actually tickle themselves!

Daniel Wolpert and his colleagues at University College, London, have also devised a simple and elegant experiment to explain the cerebellum's involvement in the escalation of a shoving match (Figure 1.2). When a shoving match starts between two people the force of the shoving tends to escalate, often to the point of a full-blown brawl. Typically, we have thought of this solely in terms of social dynamics: neither participant wants to show weakness by backing down. That may explain why the conflict continues, but it does not necessarily shed light on why the force of each shove increases in a tit-for-tat exchange.

What Wolpert and his colleagues did was have two adult subjects face each other, each resting the left index finger, palm up, in a molded depression. A small metal bar on a hinge was then rested lightly on top of each subject's finger. The hinge was fitted with a sensor to measure the force delivered when the bar was pressed down. Both subjects were given the same instructions: exactly match the force of the tap on his finger that he receives with an equivalent tap when his turn comes. Neither subject knew the instructions given to the other.

Despite explicit instructions to the contrary, when the subjects took turns pressing on each other's fingers, the force applied always escalated dramatically, just as it does in schoolyard or bar-room confrontations. Each person swore that he matched the force of the other's tap. When asked to guess the instructions given to the other person, each said, "You told the other guy to press back twice as hard."

Why does this happen? Several clues point to the answer. First, it is not specific to social situations. When a person is asked to match the force of a finger tap which comes from a machine, he or she she will also respond with greater force. The second line of evidence comes from modifying the tit-for-tat experiment so that the tap is produced not by pressing on a bar but rather by moving a joystick that controls the pressure by activating a motor. The important difference between these two situations is that when the force is generated by bar pressing, making a stronger tap requires generating more force with the fingertip. When the joystick is used, however, the motor does the work and there is only a weak correlation between the force generated by the tapping finger and the force produced on the upturned finger of the other subject. When the tit-for-tat experiment is then repeated with joysticks there is very little force escalation. The interpretation here is similar to that offered for self-tickling: The cerebellum receives a copy of the commands to produce the finger tap (using the bar) that are proportional to the force applied. It then creates a prediction of the expected sensation that is sent to the somatosensory cortex to inhibit feedback sensations from the fingertip during tapping. To overcome this inhibition, the subject presses harder to match the force perceived from the last tap he or she received, thus escalating the force applied.

So, in most situations, the cerebellar circuit that allows us to pay less attention to sensations that result from self-generated movement and more attention to the outside world is a useful mechanism. But as any 8-year-old coming home with a black eye and a tale of "But Mom, he hit me harder!" will tell you, there is a price to pay for this feature. This is a common brain design flaw. Most systems, like the cerebellar inhibition of sensations from self-generated movement, are always on. They cannot be switched off even when their action is counterproductive.

Moving up and forward from the cerebellum, the next region we encounter is called the midbrain. It contains primitive centers for vision and hearing. These locations are the main sensory centers for some animals, such as frogs or lizards. For example, the midbrain visual center is key for guiding the tongue-thrust frogs use to capture insects in flight. But in mammals, including humans, the midbrain visual centers are supplemented and to some degree supplanted by more elaborate visual regions higher up in the brain (in the cortex). Even though we make only limited use of a frog-like visual region in our brains (mostly in orienting our eyes to certain stimuli), this evolutionarily ancient structure has been retained in human brain design and this gives rise to the fascinating phenomenon called blindsight.

Patients who are effectively blind owing to damage to the higher visual parts of the brain will report that they have no visual sense whatsoever. When asked to reach for an object in their visual field, such as a penlight, they will say, "What can you possibly mean? I can't see a thing!" If however, they are told to just take a guess and try anyway, they can usually succeed at this task at a rate much higher than would be due to pure chance. In fact, some patients can grasp the penlight 99 percent of the time, yet will report each time that they have no idea where the target is and they are guessing randomly. The explanation seems to be that the ancient visual system in the midbrain is intact in these patients and guides their reaching, yet because this region is not interconnected with the higher areas of the brain, these people have no conscious awareness of the penlight's location. This underscores a general theme that is emerging here. The functions of the lower portions of the brain such as the brainstem and the midbrain are generally performed automatically, without our conscious awareness. As we continue our tour to those parts of the brain that are both literally and metaphorically higher, then we will begin to make the transition from subconscious to conscious brain function.

Furthermore, the midbrain visual system is a lovely example of brain kludge: it is an archaic system that has been retained in our brains for a highly delimited function, yet its action can be revealed in brain injury. As an analogy, imagine if your present-day audio electronics, let's say that sleek handheld MP3 player, still contained a functional, rudimentary 8-track tape player from the 1960s.

Not too many of those would get sold, even with a really urban-hip, edgy ad campaign.

Moving a bit upward and forward, we reach two structures called the thalamus and the hypothalamus (which just means "below the thalamus"). The thalamus is a large relay station for sending sensory signals on to higher brain areas and also relaying command signals from these areas out along pathways that ultimately activate muscles. The hypothalamus has many smaller parts, each of which has a separate function, but one general theme of this region is that it helps to maintain the status quo for a number of body functions, a process called homeostasis. For example, when you get too cold, your body begins to shiver reflexively in an attempt to generate heat through muscular activity. The shivering reflex originates within the hypothalamus.

Perhaps the most well-known homeostatic drives are those that control hunger and thirst. Although the urge to eat and drink can be modulated by many factors, including social circumstances, emotional state, and psychoactive drugs (consider the well-known phenomenon of "the munchies" from smoking marijuana and the appetite-suppressing action of amphetamines), the basic drives for hunger and thirst are triggered within the hypothalamus. When tiny holes are made surgically in one part of the hypothalamus of a rat (called the lateral nucleus; a "nucleus" in the brain is just a name for a group of brain cells), it will fail to eat and drink, even after many days. Conversely, destroying a different part of the hypothalamus (the ventromedial nucleus) results in massive overeating. Not surprisingly, a huge effort is under way to identify the chemical signals that trigger feelings of hunger and fullness, with the hope of making a safe and effective weight-loss drug. So far, this has proven to be much more difficult than anticipated because multiple, parallel signals for both beginning and ending feeding appear to play a role.

In addition to its involvement in homeostasis and biological rhythms, the hypothalamus is also a key controller of some basic social drives, such as sex and aggression. I will talk about these functions in detail later. A point that must be made here, though, is that the hypothalamus exerts some of its effects on these drives by secreting hormones, powerful messenger molecules that are carried in the bloodstream throughout the body to cause many varied responses. The hypothalamus secretes two types of hormones. One type has direct actions on the body (such as the hormone called vasopressin, which acts on the kidney to limit the formation of urine and thereby increase blood pressure), and the second type, the so-called master hormones, directs other glands to secrete their own hormones. A good example of the latter is growth hormone, secreted by the pituitary gland in growing children and adolescents but stimulated by a master hormone released by the hypothalamus. After much careful scientific thought, this master hormone was given the compelling name "growth hormone releasing hormone" (endocrinologists, like many scientists, are not known for their literary flair).

Up to this point, we have been looking at the brain sliced exactly down the middle. Many areas inside the brain are revealed with this view, but others are buried deep within the tissue and are not visible either from the outside surface or from the cut surface at the midline. Particularly important are two deeply buried structures called the amygdala ("almond") and the hippocampus ("seahorse") that constitute part of a larger circuit in the center of the brain called the limbic system (which also contains portions of the thalamus, cortex, and other regions). The limbic system is important for emotion and certain kinds of memory. It is also the first place in our bottom-to-the-top tour where automatic and reflexive functions begin to blend with conscious awareness.

The amygdala is a brain center for emotional processing that plays a particular role in fear and aggression. It links sensory information that has already been highly processed by the cortex (that guy in the ski mask jumping out of that dark alley at me can't be up to any good) to automatic fight-or-flight responses mediated by the hypothalamus and brainstem structures (sweating, increased heart rate, dry mouth). Humans rarely sustain damage to the amygdala alone, but those who do often have disorders of mood and appear to be unable to recognize fearful expressions in others. Electrical stimulation of the amygdala (as sometimes occurs during neurosurgery) can evoke feelings of fear, and the amygdala also appears to be involved in storing memories of fearful events.

The hippocampus (which, when dissected out of the brain, actually looks more like a ram's horn than the seahorse for which it is named) is a memory center. Like the amygdala, it receives highly processed sensory information from the cortex lying above it. Rather than mediating fear, however, the hippocampus appears to have a special role in laying down the memory traces for facts and events, which are stored in the hippocampus for a year or so but are then moved to other structures. The most compelling evidence for this model comes from a small number of people who have sustained damage to their hippocampus and some surrounding tissue on both sides of the brain. The most famous of these cases is called H.M. (initials used to protect privacy), a man who in 1953 underwent surgical removal of the hippocampus and some surrounding tissue on both sides of his brain in order to control massive seizures that had not responded to other treatments. The surgery was successful in controlling his epilepsy and did not impair his motor functions, language, or general cognitive abilities, but there were two disastrous side effects. First, H.M. lost his memory of everything that occurred 2–4 years before the surgery. He had extensive, detailed, and accurate recall of earlier events, but his memory of his life in the years just before the surgery is lost forever. Even more devastating is that since the surgery H.M. has been unable to store new memories for facts and events. If you were to meet him on Monday, he would not remember you on Tuesday. He can read the same book every day and it will be new to him. Although he has short-term memory that can span tens of minutes, his ability to store new permanent memories for facts and events is gone.

The seminal insights about memory and the hippocampus that came from H.M.'s case have since been reinforced many times, both by other patients who, for a variety of reasons, have sustained similar damage, and by animal studies in which the hippocampus has been surgically destroyed or had its function disrupted by drugs. A consistent and simple conclusion comes from this work: without a hippocampus, the ability to store new memories for facts and events is severely impaired.

Finally, moving to the outer surface of the brain, we reach the cortex. The cortex of the human brain is massive. The functions of some areas in the cortex are well understood, but others are terra incognita. A portion of the cortex analyzes the information coming from your senses. The very back of your cortex is where visual information first arrives, and another strip of tissue just behind the main sideways groove in your brain (called the central sulcus) is where touch and muscle sensation first arrives. Similar maps can be drawn for other senses. If we stimulate these areas with an electrode we can mimic activation of the sensory system involved: stimulating the primary visual cortex will cause a flash of light, or something similar, to be seen. Likewise, there is a strip of cortex just in front of the central sulcus that sends out command signals that ultimately cause contraction of muscles and consequent body movement. Electrically stimulating this motor cortex results in muscular contraction. This is a standard technique for making a functional map of the brain when surgery must be performed in this area. What's most interesting about the cortex are those regions for which the functions are not obviously either sensory or motor. Brain researchers have sometimes called these regions association cortex. Association areas are most plentiful in the front of the brain (the frontal cortex), a region that is highly developed in humans.

I have offered a number of examples where people (and experimental animals) sustain damage to various brain regions and suffer various losses of function ranging from amnesia to overeating. Yet, to this point, though many of these brain insults have had devastating effects, none of them has changed the personality, the essential core identity of the sufferer. H.M., for example, has the same unique personality that he had before his epilepsy surgery. A far different picture emerges when we consider damage to the frontal cortex.

Here, the most well known example is Phineas Gage, a foreman on a Vermont railway gang in 1848. Railway construction, then and now, uses blasting to remove obstacles and level the roadbed. Phineas, aged 25, had the unenviable task of jamming the explosive charge into place using a long metal rod known as a tamping iron. You can imagine what happened. As he stood over a borehole, tamping the charge, there was a spark that ignited a horrible explosion. The explosion drove the tamping iron through Phineas's left cheek and eye at a steep upward angle, piercing his skull through the eye socket tearing a huge hole in his left frontal cortex, and exiting his skull through the top. Figure 1.3 shows a drawing based on a scan of his skull made long after his death, with the tamping iron in place. Amazingly, after a few weeks in bed, Phineas made a full recovery. The infection of his wound abated. He could walk, talk, and do arithmetic in his head. His long-term memory was fine. What *had* changed was his personality and his judgment. By all reports, before the accident he was kind, level-headed, friendly, and charismatic. After his recovery he became arrogant, opinionated, impulsive, rude, and selfish. Not to put too fine a point on it, damage to his frontal cortex changed him from a nice guy into a jerk. His former coworkers couldn't stand him. "He's just not Gage anymore," one

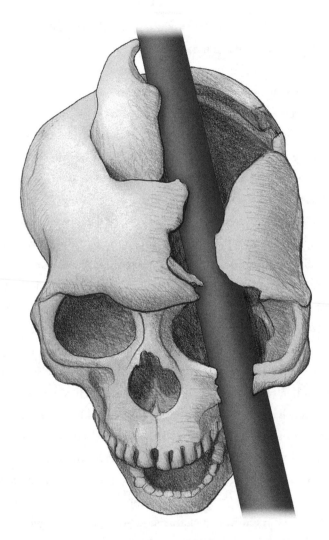

FIGURE 1.3. The skull of Phineas Gage, with the famous tamping iron, reconstructed by computer from scans made long after his death. Derived with permission from P. Ratiu and I.-F. Talos, Images in clinical medicine: the tale of Phineas Gage, digitally remastered, *The New England Journal of Medicine* 351:e21 (2004). *Joan M. K. Tycko, illustrator.*

friend reportedly said. Tragically, he ended up in a carnival freak show, reinserting the tamping iron through the healed but still present hole in his head to the morbid fascination of onlookers. He died 12 years after the tamping iron accident.

As shown by the case of Phineas Gage, and documented many times since, the frontal cortex is the substrate of our individuality, determining our social interactions, outlook, and perhaps even our moral sense. Not just our cognitive capacities but our character—our personhood, so to speak—resides in this most recently evolved region of our brains.

HAVING COMPLETED OUR whirlwind tour from the bottom to the top of the brain (leaving out a few areas), what can we conclude about the overall principles of brain design? Guiding Principle One: The highest functions of our brain, involving conscious awareness and decision making, are located at the very top and front, in the cortex, and the lowest functions, supporting basic subconscious control of our body functions such as breathing rhythm and body temperature, are located in the very bottom and rear, in the brainstem. In between are centers that are engaged in higher subconscious functions such as rudimentary sensation (midbrain), homeostasis and biological rhythms (hypothalamus), and motor coordination and sensory modulation (cerebellum). The limbic system, including the amygdala and hippocampus, is the crossroads where the conscious and unconscious parts of the brain meet and initiate the storage of certain types of memories.

Guiding Principle Two: The brain is built like an ice cream cone (and you are the top scoop): Through evolutionary time, as higher functions were added, a new scoop was placed on top, but the lower scoops were left largely unchanged. In this way, our human brainstem, cerebellum, and midbrain are not very different in overall plan from that of a frog. It's just that a frog has only rudimentary higher areas in addition (barely more than one scoop). All those structures plus the hypothalamus, thalamus, and limbic system are not that different between humans and rats (two scoops), which have a small and simple cortex, while we humans have all that plus a hugely elaborated cortex (three scoops). When new, higher functions were added, this did not result in a redesign of the whole brain from the ground up; a new scoop was just added on top. Hence, in true kludge fashion, our brains contain regions, like the midbrain visual center, that are functional remnants of our evolutionary past.

You probably have seen those quaint charts from the nineteenth century (Figure 1.4), in which the surface of the brain is divided into neat regions, each labeled with a cognitive function (such as calculation) or a personality trait (say combativeness). The phrenologists who used these charts believed not only that those functions could be mapped to those particular brain regions but also that bumps on the skull resulted from the overgrowth of a particular brain region. Indeed, there was a cottage industry in the nineteenth and early twentieth centuries of professional head-bump feelers, who, armed with charts, plaster models, and even a mechanical bump-measuring helmet, would analyze the skull-and-mind of anyone willing to pay.

The phrenologists were wrong on two counts. First, bumps on the skull don't indicate anything about the underlying brain tissue. Second, their diagrams equating particular regions with cognitive functions and personality traits were pure fantasy. But on a more general issue, the phrenologists were right: the brain is not an undifferentiated mass of tissue where each region contributes equally to all functions. Rather, particular brain functions often are localized to distinct brain regions.

This brings us to Guiding Principle Three: Localization of function in the brain is straightforward for basic subconscious reflexes such as vomiting and is fairly straightforward for the initial stages of sensation (we know where signals

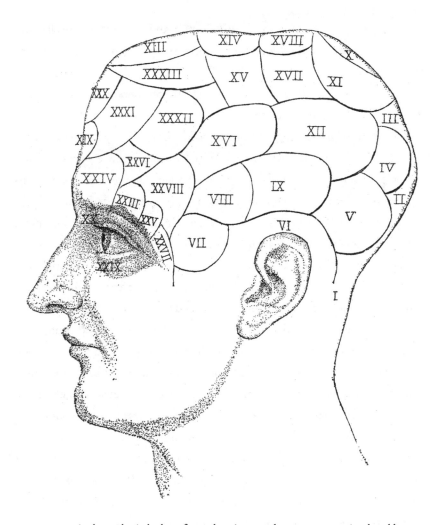

FIGURE 1.4. A phrenologist's chart from the nineteenth century, equating head bumps with particular mental traits. In this case, for example, XIV = veneration, XVII = hope, XIII = benevolence, XXI = imitation, XIX = ideality, VIII = acquisitiveness, XVIII = marvelousness, and XX = wit. *From W. Mattieu Williams, A Vindication of Phrenology (Chatto & Windus, London, 1894).*

first arrive in the cortex for vision, hearing, smell, and so forth) But localization of function is much more difficult for more complex phenomena such as memory of facts and events and is really hard for the highest functions such as decision making. In some cases it becomes complicated because the location of a function in the brain is not fixed over time: memories for facts and events seem to be stored in the hippocampus and some immediately adjacent regions for 1–2 years but are then exported to other locations in the cortex. Decision making generally is such a broad function, and generally requires such a convergence of information, that it may be broken into smaller tasks and distributed to a number of places in the cortex. We may have to define functions more precisely in order to achieve a greater understanding of functional localization.

So, given these Guiding Principles, what is it about this organ that makes us so clever? What is it about our brain that enables language and the ability to understand the motivations of others (the so-called theory of mind) and other capacities that humans have developed far beyond the abilities of other animals? We don't have the biggest brains (an elephant's is bigger) and we don't even have the biggest brain-to-body-weight ratio (small birds beat us on that measure). We don't have the most wrinkled brain surface (whales and dolphins' are more wrinkled). In fact, we don't even have the largest brains among our hominid kin: estimates derived from skull volumes indicate that Neanderthals had brains that were, on average, somewhat larger than ours today. And, although I haven't talked about it yet, we can assume that, overall, the shape and chemical composition of the cells that make up our brains are not fundamentally different from those of a rat (more on this to come). What we do have is the largest association cortex, that which is not strictly sensory or motor, most of it packed into the front half of our brain. Somehow, this is the elaboration that appears to have given humans their cognitive advantages.

Can we take this one step farther? Humans have varying cognitive abilities.

Can human cognitive capacity be predicted by the overall size of the brain or by the size of particular brain regions? Diseases (both inherited and acquired) and trauma, both of which produce gross anatomical disruptions to the brain, can clearly impair cognition. But what about normal variation, excluding obvious mishaps such as trauma or disease? Recent studies relating normal human variation in cognitive ability to brain size or shape have used brain-scanning techniques that provide more accurate measures than older studies that relied upon skull measurements. In general, these newer studies have found statistically significant correlations between brain size (adjusted for body weight) and cognitive ability. But this correlation, while real, accounts for only about 40 percent of the variation in cognitive ability of normal humans. Thus one can find people at the small end of the range of normal brain sizes (say, 1,000 cubic centimeters) who will score highly on a so-called test of general intelligence. Conversely, one can find individuals with unusually large brains (1,800 cubic centimeters) who score well below average.

The large variation in the relationship between human brain size or shape and cognitive capacity has not stopped the continual trickle of publications in which the preserved brains of famous historical figures have been analyzed anatomically. Lenin's brain was studied in Germany in the late 1920s and, while it was of average weight, in some regions a particular subset of cells in the brain (called layer 3 cortical pyramidal cells) were purported to be unusually large compared to other postmortem samples. Einstein's brain actually was smaller than average (but well within the normal range). Recently, there has been a claim that a region of his brain called the inferior parietal cortex was slightly (15 percent) enlarged relative to a sample of men's brains of a similar age. That caused some interest because this region has been associated with spatial and mathematical cognition, areas in which Einstein clearly excelled. But one must be cautious in interpreting this sort of finding. First, it's very hard to make a claim based on a single sample (Einstein). A more convincing study would need a whole group of mathematical/spatial geniuses compared with controls carefully matched for age, lifestyle, and other factors. Second, and more important, there's a problem of causality at work. If, indeed, a part of Einstein's brain involved in mathematical/spatial thinking was significantly larger than appropriate control brains, does that mean that this variation endowed him with mathematical ability that he was then able to exploit? Or did his lifelong engagement in mathematical and spatial pursuits cause this part of his brain to grow slightly?

Failure up to now to strongly associate gross anatomical features of the brain with normal variation in human cognition should not be taken to mean that variation in human cognition has no measurable physical correlate in brain structure. It's very likely that such a relationship does exist. But this correlation will be only weakly reflected in crude measures such as brain size. Most of human cognitive variation is more likely to be manifest as changes in the microscopic anatomy, the connectivity of brain cells, and the patterns of brain electrical activity.

WE'VE UNCOVERED THREE Guiding Principles of Brain Design and these highlight a few of the ways in which the human brain is poorly organized. The brain has primitive systems that developed in our distant evolutionary past (before mammals) and that have been supplemented by newer, more powerful structures. These primitive structures persist in the lower parts of our brain, giving rise to interesting phenomena such as blindsight. Also, the brain has regions that perform functions that are often useful, such as cerebellar inhibition of the sensations from self-originated movements, but that cannot be turned off in the appropriate circumstances, a fact that contributes to problems such as force escalation in tit-for-tat conflicts.

To put this in perspective, imagine that you are an engineer in charge of building the latest and most efficient car. Only after you agree to take the job do you learn that there are two weird stipulations. First, you are given a 1925 Model T Ford and told that your new car must take the form of adding parts to the existing structure while taking almost nothing of the original design away. Second, most of the new complex control systems you will build, such as the device that rapidly pumps the antilock brakes, must remain on all of the time (not just when a skid is detected). These are some of the types of constraints that have influenced the design of the human brain as it has evolved. Together with the engineering flaws of the component parts (the cells of the brain, which I will consider in Chapter 2) and the assembly process (brain development, covered in Chapter 3), these aspects of suboptimal design are central to brain function. By the end of this book I hope to have convinced you that almost every aspect of transcendent human experience, including love, memory, dreams, and even our predisposition for religious thought, ultimately derives from the inefficient and bizarre brain engineered by evolutionary history.

Chapter Two

Building a Brain with Yesterday's Parts

IT IS A CLICHÉ to be awed by the microscopic complexity of the human brain. Any scientist who talks about this topic inevitably hears the kindly, avuncular ghost of Carl Sagan whispering: "Bill-yuns and bill-yuns of tiny brain cells!" Well, it is rather impressive. There are a hell of a lot of cells in there. The two main cell types in the brain are: neurons, responsible for rapid electrical signaling (the brain's main business), and glial cells, important for housekeeping functions that create an optimal environment for neurons (and that directly participate in some forms of electrical signaling as well). The famous numbers: approximately 100 billion (100,000,000,000) neurons in the adult human brain and approximately one trillion (1,000,000,000,000) glial cells. To put this in perspective, if you wanted to give your neurons away to all humanity, everyone on earth would receive about 16 of them.

Neurons are not a recent development in evolution. They are soft and therefore not well preserved in fossils, so we don't know exactly when the first neurons appeared. But we do know that modern jellyfish, worms, and snails all have neurons. Some other modern animals, such as sea sponges, don't. Therefore, our best guess is that neurons appeared at about the time when jellyfish and their relatives, a group of animals called *Cnidaria,* first appeared in the fossil record, in the Pre-Cambrian era, about 600 million years ago. Incredibly, with few exceptions, the neurons and glial cells in a worm are not substantially different from those in our own brains. In this chapter, I hope to show you that our brain cells have an ancient design that makes them unreliable and slow, and limits signaling capacity.

Neurons come in a variety of shapes and sizes (see Figure 2.1), but have certain structures in common. Like all cells, neurons are bounded externally by a sort of skin, the outer membrane (also called the plasma membrane). All neurons have a cell body, which contains the cell nucleus, the storehouse of genetic instructions encoded in DNA. The cell body can be round, triangular, or spindle shaped and can range from 4 to 100 microns across (20 microns is typical). Perhaps a more useful way to think about this is that five average-sized neuronal cell bodies could be placed side by side in the width of a typical human hair. Thus the outer membranes of neurons and glial cells are incredibly tightly packed with very little space in between.

Sprouting from the cell body are dendrites (from the Greek word for "tree"), large, tapering branches of the neuron that receive chemical signals from neighboring neurons. I'll discuss how this happens soon. Dendrites can be short or long, spindly or bushy or, rarely, even completely absent. High magnification shows that some are smooth while others are covered with tiny nubbins called dendritic spines. Typical neurons have several branching dendrites, but they also have a single long thin protrusion growing from the cell body. This is the

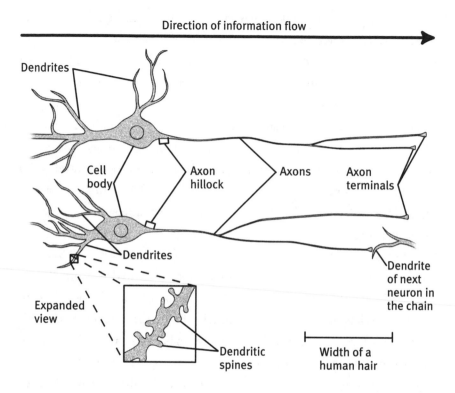

FIGURE 2.1. Two different neurons with their parts labeled. *Joan M. K. Tycko, illustrator.*

axon and is the information-sending side of the neuron. The axon, usually thinner than the dendrites, does not taper as it extends from the cell body. A single axon grows from the cell body, but it often subsequently branches, sometimes going to very different destinations. Axons can be remarkably long: some run all the way from the base of the spine to the toes (which makes the longest axons around 3 feet for average humans, and up to 12 feet long for a giraffe).

At specialized junctions called synapses, information passes from the axon of one neuron to the dendrite (or sometimes the cell body) of the next (Figure 2.2).

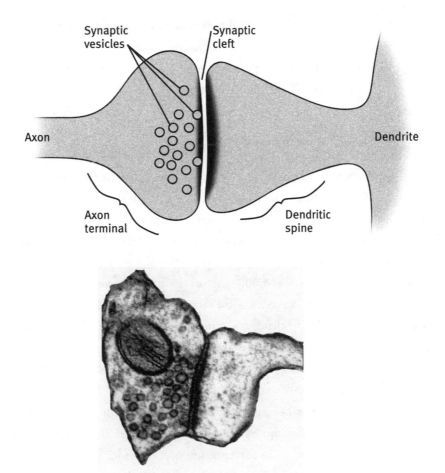

Synaptic vesicles

Synaptic cleft

Axon

Dendrite

Axon terminal

Dendritic spine

FIGURE 2.2. Parts of the synapse in a drawing (top) and in an actual electron microscope photo (bottom). *Joan M. K. Tycko illustrated the top panel. The bottom panel was kindly provided by Professor Kristen Harris of the Medical College of Georgia. Her website, synapses.mcg.edu, provides an excellent overview of the fine structure of synapses.*

At synapses, the ends of axons (called axon terminals) nearly, but not actually, touch the next neuron. Axon terminals contain many synaptic vesicles, tiny balls with a skin made of membrane. The most common type of synaptic vesicle in the brain is loaded with about 2,000 molecules of a specialized compound called a neurotransmitter. Between the axon terminal of one neuron and the dendrite of the next is a tiny saltwater-filled gap called the synaptic cleft. By tiny, I mean extremely tiny: about 5,000 synaptic clefts would fit in the width of a single human hair. The synaptic cleft is the location where synaptic vesicles release neurotransmitters to signal the next neuron in the chain.

Synapses are crucial to our story. They will come up repeatedly as I discuss everything from memory to emotion to sleep. We should therefore spend some time on them now. First, the number of synapses in the brain is staggering. On average, each neuron receives 5,000 synapses, locations where the axon terminals of other neurons make contact (the range is from 0 to 200,000 synapses). Most synapses contact the dendrites, some the cell body, and a few the axon. Multiplying 5,000 synapses per neuron by 100 billion neurons per brain, gives you an estimate of the astonishing number of synapses in the brain: 500 trillion, 500,000,000,000,000.

Synapses are the key switching points between the two forms of rapid signaling in the brain: chemical and electrical impulses. Electrical signaling uses a rapid blip, called a spike, as its fundamental unit of information. Spikes are brief electrical signals that originate at the axon hillock, the place where the cell body and the axon join. When spikes, having traveled down the axon, arrive at the axon terminals they trigger a series of chemical reactions that cause a dramatic structural change (see Figure 2.3). Synaptic vesicles fuse with the outer membrane of the axon terminal, dumping their contents, special neurotransmitter molecules, into the synaptic cleft. These neurotransmitter molecules then move across the synaptic cleft, where they contact specialized proteins

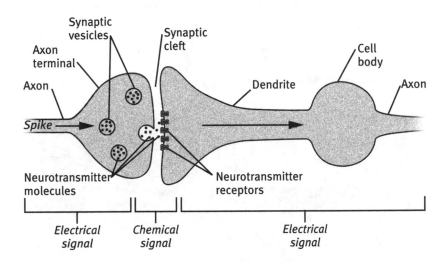

FIGURE 2.3. Synapses, the key sites in the brain for converting electrical signals to chemical signals and then back into electrical signals. Reading from left to right tells the story of synaptic signaling. *Joan M. K. Tycko, illustrator.*

called neurotransmitter receptors, embedded in the membrane of a neighboring neuron's dendrite. Receptors convert the neurotransmitter's chemical signal back into an electrical signal. Electrical signals from activated receptors all over the dendrite are funneled toward the cell body. If enough electrical signals arrive together, a new spike is triggered and the signal is passed farther along the chain of neurons.

That's the *Reader's Digest* version. Now, let's flesh that out with some real biology. At about 3 pounds, the brain constitutes about 2 percent of total body weight, and yet it uses about 20 percent of the body's energy. Clearly, the brain is an inefficient energy hog (the Hummer H2 of the body, if you will), but why is this so? The brain is naturally bathed in a special saltwater solution called cerebrospinal fluid that has a high concentration of sodium and a much lower

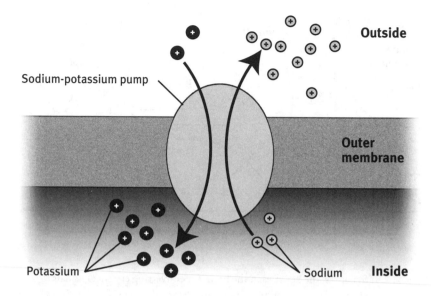

Outside

Sodium-potassium pump

Outer membrane

Potassium

Sodium

Inside

FIGURE 2.4. The sodium-potassium pump. Located in the outer membrane of neurons, it pumps sodium ions out and potassium in, thereby establishing the electrical gradient used by neurons to send information. *Joan M. K. Tycko, illustrator.*

concentration of potassium. These sodium and potassium atoms are in their charged state, called ions, in which they each have one unit of positive charge (+1). The brain's main energy expense involves continuously running a molecular machine that pumps sodium ions out of the cell and potassium ions in (see Figure 2.4). As a result of this pump's action, the concentration of sodium ions outside a neuron is about 10-fold higher than it is inside. For potassium, the concentration gradient runs in the other direction: the concentration of potassium ions is about 40-fold greater inside than outside. So neurons have saltwater solutions on both sides of their outer membranes (the skin of the cell), but very different saltwater solutions: the outside solution is high in sodium and low in potassium; the inside solution is the opposite, low in sodium and high in potassium. That is the basis of electrical function in the brain. The differences in concentrations of sodium and potassium create potential energy, similar to that created by winding the spring on a child's toy, that can then be released in the appropriate circumstances to generate neural signals. Neurons rest with an electrical potential across their outer membranes: there is more negative charge inside the cell than outside.

Let's conduct an imaginary experiment that will help us understand neuronal electrical signaling. In our imagined lab, some neurons have been extracted from a rat's brain, placed in petri dishes, and grown in special solutions designed to mimic cerebrospinal fluid. This process is called neuronal cell culture and is a standard technique in brain research laboratories. In this experiment, illustrated in Figure 2.5, we insert recording electrodes into a neuron to measure the electrical signals across the outer membrane. Recording electrodes are hollow glass needles with very fine points, filled with a special saltwater solution that mimics the neuron's internal milieu (high potassium, low sodium). One electrode is in the dendrite, where a particular synapse is received, another is at the axon hillock, the place where the axon just starts to grow from the cell body, and a third electrode is way down in the axon terminal. Yet another electrode is used, not for recording, but rather for electrical stimulation of an axon terminal of another neuron that is contacting the dendrite of the first.

Before anything happens, we record the previously mentioned negative resting potential across the outer membrane of the information-receiving neuron. Measured in thousandths of a volt, or millivolts, our typical neuron's resting potential across its outer membrane is −70 millivolts, or about 1/20th the voltage of a single AA battery. Next, we electrically stimulate the adjacent axon terminal, causing it to release neurotransmitter molecules into the synaptic cleft. In our imaginary experiment, this neurotransmitter is the molecule glutamate.

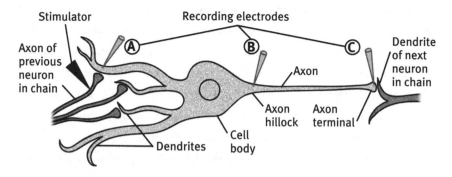

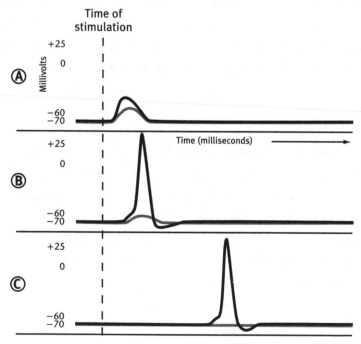

I have chosen glutamate as our example because it is by far the most common neurotransmitter molecule in the brain. When glutamate molecules are released at synapses, they diffuse across the narrow synaptic cleft separating two neurons. Glutamate molecules are not squirted across the synapse with force; they merely diffuse, like a single drop of red wine slowly mixing into a full glass of water. Because the synaptic cleft is so small, in only about 50 one-millionths of a second (5 microseconds) glutamate molecules released from the presynaptic axon terminal of one neuron cross to the other side, the postsynaptic membrane of the dendrite. Most of the glutamate molecules simply diffuse away and have no effect, but some will bind specialized glutamate receptor proteins that are embedded in the postsynaptic membrane. There are many different neurotransmitters in the brain, and though glutamate is the most common one, many others are important and will arise as I consider particular brain functions.

Glutamate receptor proteins are highly complex molecular machines. They are built of four similar parts that join together to form a doughnut-shaped structure around a central pore (Figure 2.6). In the resting state, this pore is shut tight, but when glutamate binds this receptor a gate that normally closes

FIGURE 2.5. An imaginary experiment to investigate electrical signaling in neurons. Weak stimulation (of a few terminals) gives rise to the release of glutamate molecules, which diffuse across the synaptic cleft and bind glutamate receptors to evoke the responses indicated with gray lines in the chart at the bottom of the illustration. A small excitatory postsynaptic potential (EPSP) in the dendrites is even smaller in the axon hillock and fails to trigger a spike. Strong stimulation of terminals (responses indicated with black lines) causes a large EPSP in the dendrite. This EPSP is smaller in the axon hillock but is still big enough to cause a spike to be initiated here, and this spike then travels down the axon, where, after a delay, it is also recorded in the axon terminals. *Joan M. K. Tycko, illustrator.*

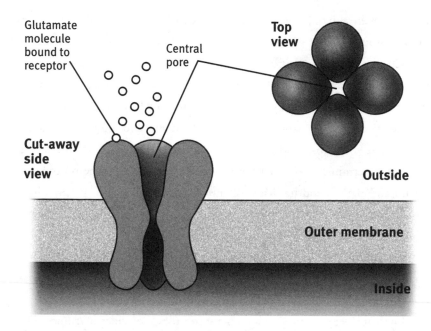

Glutamate molecule bound to receptor

Central pore

Top view

Cut-away side view

Outside

Outer membrane

Inside

FIGURE 2.6. Schematic drawing of a glutamate receptor in the postsynaptic membrane. Glutamate binding to its receptor opens the central pore, the ion channel. *Joan M. K. Tycko, illustrator.*

off this central pore opens, thus allowing certain ions to flow in or out of the cell. The receptor's central pore is small and its particular chemical properties ensure that only particular ions can get through. Hence, the central pore has been given a special name, ion channel. In the case of the glutamate receptor, the ion channel allows passage of both sodium ions and potassium ions. When the pore opens, sodium ions from the outside (where sodium concentration is high) rush to the inside (where sodium concentration is low), and potassium ions flow in the opposite direction from inside (where concentration is high) to outside (where it is low). In this process, more sodium ions rush in than potassium ions flow out, so there is a net flow of positive charge into the cell, raising the voltage difference across the dendrite's outer membrane (the membrane potential) from its resting state of -70 millivolts to some more positive level, let's say -65 millivolts. As the glutamate molecules diffuse away from their receptors and the receptor-gated ion channel (central pore) closes again, the membrane potential returns to the resting state. This whole event, about 10 milliseconds in duration from start to finish, has been given a rather long and ponderous name, the excitatory postsynaptic potential, abbreviated EPSP.

In most neurons, a single EPSP produces a response like the one we have seen, a brief change in voltage, then nothing. This is a fairly typical mechanism that neurons have for ignoring very low levels of activity that are merely ongoing noise in the brain. Something very different happens if we activate a group of axon terminals to release glutamate all at the same time. We produce a larger EPSP at both the dendrite and the axon hillock, but when the strength of the signal at the axon hillock reaches a certain threshold level (say about -60 millivolts), an amazing thing happens. Rather than falling back down to rest, the membrane potential at the axon hillock explosively deflects upward and then rapidly returns. This explosive response is the spike, the fundamental unit of information in the brain.

Why is there a spike and why does it start at the axon hillock? The answer is in the structure of the outer membrane at this location. The axon hillock, but not the dendrite or cell body, has a high density of a different ion channel. These ion channels are not opened by binding glutamate, but rather have a built-in sensor of the local membrane voltage that allows them to be shut at rest (-70 millivolts) but open when the membrane voltage becomes more positive (to about -60 millivolts and beyond). When EPSPs from several different synapses add up at the axon hillock and move the membrane potential to -60 millivolts, then these voltage-sensitive ion channels begin to open. They are built to allow only sodium ions through their central pore, and as this sodium rushes in, it moves the membrane to an even more positive potential. This, in turn, causes more voltage-sensitive sodium channel opening in a rapid positive feedback loop that underlies the explosive upstroke of the spike.

The spike typically peaks at about +50 millivolts and rapidly falls back to rest. There are two factors that contribute to this rapid peak-and-return behavior. First, voltage-sensitive sodium ion channels open rapidly but stay open only for about a millisecond before snapping closed again, which limits the spike's duration. Second, there is another type of voltage-sensitive ion channel involved. This one is also activated by positive-going changes in membrane potential, but it opens more slowly and when it opens, potassium ions rush out of the neuron. The loss of positively charged potassium ions from inside the cell makes the membrane potential more negative, causing the downstroke of the spike as the membrane potential returns to rest.

The axon hillock, where the spike originates, is the first stretch of a long highway to the axon terminal. Fortunately, the voltage-sensitive sodium channel's positive feedback loop allows the spike to travel along the axon. Sodium ions rushing in make the outer membrane more positive not just at the axon hillock, but also at the next bit of axon, farther from the cell body. Because the membrane in this next bit of axon also has voltage-gated sodium channels, they will open, sodium ions will rush in at that location and produce more positive charge in yet a further bit of axon membrane, and so on. In this manner, the spike travels down the axon like a flame racing along a fuse, each bit of axonal membrane "igniting" the next until the spike reaches the axon terminals.

The voltage-sensitive sodium channel that initiates neuronal spikes is a key target of neurotoxins generated by many plants and animals. Interfere with that channel and you block essentially all signaling in the brain (and the rest of the nervous system too). The most famous—or infamous—toxin is that of the

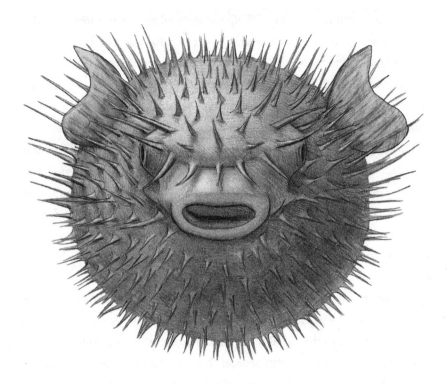

FIGURE 2.7. The pufferfish. *Joan M. K. Tycko, illustrator.*

fugu, the Japanese pufferfish (Figure 2.7). This toxin (called tetrodotoxin) is a tiny molecular plug that fits exactly into the outer portion of the sodium channel's central pore, thereby stopping it up. Tetrodotoxin is more than 1,000 times as powerful as cyanide and a single pufferfish has enough to kill 30 people. Considered a delicacy in Japan, pufferfish killed many people before preparation of fugu in restaurants was closely regulated by law to prevent people from ingesting the parts of the fish that have the highest concentrations of the toxin. Even today fugu is the one food the emperor and his family are prohibited from eating.

But let's return to movement along the axon when it is not interrupted by neurotoxins or other means. It is tempting to say that the axon is like an insulated copper electrical wire. But this obscures one of the fundamental inefficiencies of neurons. Copper wire need not do anything to keep electrical signals moving: it is totally passive, is a good conductor, and is well insulated against losing electrical charge to the outside. As a consequence, electrical signals in copper wires move at nearly the speed of light, about 669 million miles per hour. In contrast, the axon uses molecular machines with moving parts (voltage-sensitive ion channels snapping open and closed) to maintain the spike as it travels down its pathway. Comparatively, the axon is a quite poor conductor. The saltwater solution on the inside of the axon is not nearly as good a conductor as copper. Moreover, the outer membrane of the axon is a rather leaky insulator.

Perhaps the conduction of electrical signals along the axon is best understood through a hydraulic analogy. Insulated copper wire is like a steel water pipe (does not leak) that is 10 feet in diameter (great flow through its core), while the axon is like a "soaker" garden hose, 1 inch in diameter (poor flow through its core), that has been riddled with tiny holes along its length (leaks like hell) to allow you to irrigate a flower bed. This combination of poor core flow and leakiness makes water flow through a soaker hose slowly. Similarly, electrical current flow through an axon is also restricted by poor core flow and leakiness. As a consequence, electrical signals in axons typically travel slowly, at about 100 miles per hour. There is, however, quite a range, with the thinnest, uninsulated axons poking along at about 1 mile per hour and the very fastest (thick axons or those well insulated by neighboring glial cells) going at about 400 miles per hour. Nonetheless, even the very fastest axons, like those involved in reflexively withdrawing your finger from a hot stove, are conducting electrical signals at less than one-millionth the speed of copper wires.

Another way that our neurons differ from man-made devices, such as computers, to which they are often compared, involves the temporal range of their signals. The pattern of spike firing is the main way neurons encode and convey information, so timing limits on spike firing are particularly important. A desktop computer's central processing unit (circa 2006) may conduct 10 billion operations per second, but a typical neuron in a human brain is limited to around 400 spikes per second (though some special neurons, such as those in the auditory system that encode high-frequency sound, can fire up to 1,200 spikes per second). Furthermore, most neurons cannot sustain these highest rates for long (more than a few seconds) before they need a rest. With such constraints on speed and timing, it seems amazing that the brain can do what it does.

TO RETURN TO our neuronal story, we last left the spike racing down the axon highway to meet its fate. When the spike reaches the axon terminal it produces its characteristic explosive positive deflection in membrane potential. But, in the terminal, in addition to causing voltage-sensitive sodium channels to open, this voltage change also opens another class of ion channels that selectively pass calcium ions. Like sodium ions, calcium ions are positively charged (they have a charge of $+2$) and have a much higher concentration outside the cell than inside. So, like sodium ions, they too rush inside when a calcium channel is opened.

When calcium ions rush into the terminal they not only produce positive deflection in membrane potential, but also trigger unique biochemical events. Special sensor proteins for calcium ions are built into the neurotransmitter-containing synaptic vesicles. These sensors, upon binding calcium ions, set in motion a complex biochemical cascade that results in the presynaptic vesicle contacting a specialized patch of membrane called the release site and then fusing with it. Fusion of a vesicle causes the formation of a structure that resembles the Greek capital letter omega (Ω), which allows the contents of the vesicle, the glutamate molecules, to diffuse into the synaptic cleft and ultimately bind postsynaptic receptors (see Figure 2.3). In this way, the cycle of neuronal signaling from EPSP to spike to glutamate release to EPSP is completed and information is conveyed from neuron to neuron.

ALBERT EINSTEIN, in an oft-quoted critique of Werner Heisenberg's Uncertainty Principle, said, "God does not play dice with the Universe." By the standards of modern physics, Einstein turned out to be wrong. If I were to make the related statement "Our brains do not play dice with our synapses," it would also be wrong. At most synapses in the brain, when a spike invades the presynaptic axon terminal and causes influx of calcium ions, this does not necessarily result in vesicle fusion and the release of neurotransmitter. It is, quite simply, a matter of chance. The probability of neurotransmitter release for a single spike might be 30 percent at an average synapse in the brain. Some synapses have release probabilities as low as 10 percent and a few release neurotransmitter every single time (100 percent probability), but these are the exceptions, not the rule. Most synapses in our brains do not function reliably: rather, they are probabilistic devices.

OUR IMAGINARY EXPERIMENT has now revealed the entire cycle of electrical signaling in neurons. This is a basic template that can be used to understand many brain phenomena. But the situation is a bit more complicated than shown by just this one example. Glutamate opens an ion channel that lets positive charge into the cell. This tends to move the membrane potential in a positive direction, close to the level where a spike will fire, referred to as excitation (as in *excitatory* postsynaptic potential, EPSP). There are other neurotransmitters that produce the opposite effect, inhibition, where the probability of the postsynaptic cell's firing a spike is reduced. For example, the major inhibitory neurotransmitter in the brain is gamma-aminobutyric acid, abbreviated GABA. GABA binds a receptor that opens a channel that lets chloride ions flow into the postsynaptic neuron. Chloride ions have a negative charge (-1), and thus make the membrane potential more negative. This, not surprisingly, is called an inhibitory postsynaptic potential, or IPSP, and makes it even harder for the postsynaptic neuron to fire a spike.

In practice, whether or not a neuron fires a spike at any given moment is determined by the simultaneous action of *many* synapses, with excitatory and inhibitory actions summed to produce the total effect. Recall that the average neuron in the brain receives 5,000 synapses. Of these, about 4,500 will be excitatory and 500 will be inhibitory. Although only a small number are likely to be active at any one time, most neurons will not be driven to fire a spike from the brief action of a single excitatory synapse, but will require the simultaneous action of about 5 to 20 synapses (or even more in some neurons).

Glutamate and GABA are fast-acting neurotransmitters: when they bind their receptors, the electrical changes they produce occur within a few milliseconds. They are the dominant fast neurotransmitters in brain, but there are some other fast ones. Glycine is an inhibitory neurotransmitter that acts like GABA: it opens a receptor-associated ion channel to let chloride ions rush in and inhibit the postsynaptic neuron. The poison strychnine, which figures prominently in mystery novels, blocks glycine receptors and prevents their activation. Another example is acetylcholine, an excitatory neurotransmitter that, like glutamate, opens an ion channel that lets both sodium rush in and potassium out. This occurs in some parts of the brain, as well as at the synapses between neurons and muscles. The South American hunting arrow poison called curare blocks this receptor. Animals shot with a curare-tipped arrow become totally limp as commands from the nerves fail to activate muscular contraction.

In addition to the fast neurotransmitters, such as glutamate, GABA, glycine, and acetylcholine, there are also other neurotransmitters that act more slowly. These neurotransmitters bind a different class of receptors. Instead of opening ion channels, they activate biochemical processes inside the neurons. These biochemical events produce changes that are slow to start but that have a long duration: typically, from 200 milliseconds to 10 seconds. Many of these slow-acting neurotransmitters do not produce a direct electrical effect: the membrane potential does not change in either the positive or the negative direction after they bind their receptor. Rather, they change the electrical properties of the cell in ways that are only apparent when fast neurotransmitters also act. For example, the slow-acting neurotransmitter called noradrenaline can change the voltage at which a spike will be triggered from its normal level of −60 millivolts to −65 millivolts. In a neuron that is silent, there won't be any difference after noradrenaline release, but when that neuron receives fast synaptic input, there will be. If glutamate is released onto this neuron from synapses and this changes its membrane potential from the resting state of −70 millivolts to −65 millivolts, this will now result in a spike. This same action of glutamate in the absence of noradrenaline would fail to trigger a spike. In biochemical terms, we would say that noradrenaline has a modulatory action on spike firing: it doesn't directly cause spike firing but it changes the properties of spike firing produced by other neurotransmitters. The bottom line here is that fast neurotransmitters are suited to conveying a certain class of information that requires rapid signals, while slow neurotransmitters are better at setting the overall tone and range.

WHEN NEUROTRANSMITTERS are released into the synaptic cleft, they eventually diffuse away, achieving a low concentration. A while back, I invoked the image of a single drop of red wine released into a full water glass that, eventually, will turn the contents of the glass a very pale pink. This is fine if neurotransmitters were released only once. But, over time, if neurotransmitter molecules are repeatedly released, there must be some mechanism to clear the neurotransmitter from the cerebrospinal fluid surrounding brain cells before it achieves dangerously high concentrations (continuous activation of neurotransmitter receptors can often kill neurons). In terms of our wine glass image, with repeated drops the wine glass would eventually turn a uniform shade of pink and then red.

Essentially, when it comes to cleaning up after neurotransmitter release, someone has to take out the trash. For some neurotransmitters, there is the quintessentially American solution: burn that junk in the front yard. For example, acetylcholine is destroyed in the synaptic cleft by an enzyme specifically built for that purpose. Most other neurotransmitters get the European treatment: they are recycled. Glutamate molecules, through the actions of specialized transporter proteins in the outer membrane, are taken up into glial cells, where they undergo some biochemical processing before being sent to neurons for re-use. Most of the slow-acting neurotransmitters, such as dopamine and noradrenaline, are taken up right back into axon terminals, where they can be repackaged into vesicles and used again. Interestingly, GABA seems to go both ways: it is taken up by both axon terminals and glial cells. Some neurotransmitter transporters make excellent targets for psychoactive drugs (such as the antidepressant Prozac and its relatives) because blocking them will cause neurotransmitters in the synapse to linger and achieve higher concentrations.

ALL THE INFORMATION in your brain, from the sensation of smelling a rose, to the commands moving your arm to shoot pool, to that dream about going to school naked, are encoded by spike firing in a sea of neurons, densely interconnected by synapses. Now that we have gained an overall understanding of electrical signaling in the brain, let's consider the challenges the brain must confront as it tries to create mental function using a collection of less-than-optimal parts. The first challenge is the limitation on the rate of spike firing caused by the time it takes for voltage-sensitive sodium and potassium ions to open and close. As a result, individual neurons are typically limited to a maximal firing rate of about 400 spikes/second (compared with 10 billion operations/second for a modern desktop computer). The second challenge is that axons are slow, leaky electrical conductors that typically propagate spikes at a relatively sedate 100 miles per hour (compared with electrical signals in a man-made electronic device moving at around 669 million miles per hour). The third challenge is that once spikes have made it to the synaptic terminal, there is a high probability (about 70 percent on average) that the whole trip will have been in vain, and no neurotransmitters will be released. What a bum deal! These constraints may have been tolerable for the simple problems solved by the nervous system of a worm or a jellyfish, but for the human brain, the constraints imposed by (ancient) neuronal electrical function are considerable.

How does the brain manage to create human mental function with neurons that are such crummy parts? More to the point, given the comparisons above, how is it that our brains can easily accomplish certain tasks that typically baffle electronic computers—for example, recognizing instantly that an image of a Rottweiler taken from the front and another of a teacup poodle taken from the rear should both be classified as "dog"? This is a deep question, central to neurobiology, for which a detailed answer is not at hand. Yet a more general explanation appears to be as follows. Individual neurons are horribly slow, unreliable, and inefficient processors. But the brain is an agglomeration of 100 billion of these suboptimal processors, massively interconnected by 500 trillion synapses. As a result, the brain can solve difficult problems by using the simultaneous processing and subsequent integration of large numbers of neurons. The brain is a kludge in which an enormous number of interconnected processors can function impressively even when each individual processor is severely limited.

In addition, while the overall wiring diagram of the brain is laid down in the genetic code, the fine-scale wiring of the brain is guided by patterns of activity, which allows the strength and pattern of synaptic connections to be molded by experience, a process called synaptic plasticity (which I will consider in Chapters 3 and 5). It is the massively interconnected parallel architecture of the brain combined with the capacity for subtle rewiring that allows the brain to build such an impressive device from such crummy parts.

Chapter Three

Some Assembly Required

IT'S A DAUNTING task to develop a brain. The nervous system must be precisely constructed as the fertilized ovum develops into the mature organism. The tiny roundworm called *Caenorhabditis elegans* generates, arranges, and wires together a neural circuit of exactly 302 neurons and about 7,800 synapses. These 302 neurons must be derived from rapidly dividing precursor cells, migrate to the appropriate location in the body of the worm, and express the right proteins to make neurotransmitters, and form ion channels, receptors, and the like. Finally, these neurons must grow their axons and dendrites in the correct way to wire the whole thing together properly. Faults in creating this neural circuitry result in worms that can't wriggle properly through the soil or have problems finding food or avoiding dangerous conditions. It's a complicated recipe to specify all of these neuronal properties and connections. Fortunately, the

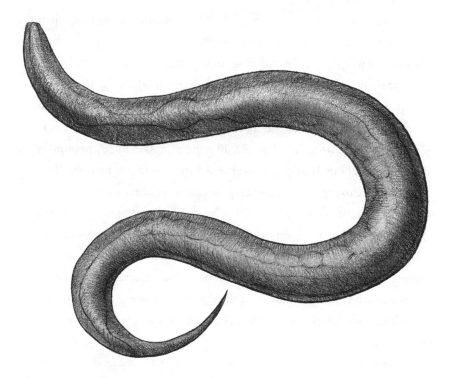

FIGURE 3.1. The roundworm *Caenorhabditis elegans,* about 1 millimeter long. It has a transparent body that allows researchers to see internal structures, including all of its 302 neurons. *Joan M. K. Tycko, illustrator.*

roundworm has, encoded in its DNA, about 19,000 genes that can potentially help guide this process.

The human brain obviously is a much bigger challenge. Its development must correctly specify the location, properties, and connections of about 100 billion neurons and about 500 trillion synapses. If all of this process were encoded in our DNA, we might expect that we would need many more genes than the roundworm. Actually, the best estimates to date from the Human Genome Project are that we have about 23,000 genes, not many more than the worm. In the human about 70 percent of these genes are expressed in the brain (the brain is not only the "energy hog" of the body, it's also a "gene hog"). Because worm neurons are really tiny and sparse and are therefore hard to dissect and analyze, we don't actually know what fraction of genes is expressed in the worm nervous system, but a reasonable guess would be 50 percent. So, as a ballpark estimate, the worm has about 9,000 genes expressed in 302 neurons while humans have about 16,000 genes expressed in 100 billion neurons. There is some evidence to suggest that human genes are more likely than worm genes to use a trick known as alternative splicing by which a single gene can give rise to multiple, related gene products. But even if we imagine that human neural genes are, on average, three times as likely to undergo splicing as their worm counterparts, we still wind up with a situation where the number of gene products per neuron (a rough measure of the capacity of genetic information to instruct brain development) is around 100 million-fold lower for humans than for worms.

So, how can our genes rise to the task at hand? How can they specify the complete development of such a large, complex structure as the human brain? The answer, simply, is that they can't: although the overall size and shape of the brain and the large-scale pattern of connections between brain regions and cell types are instructed by genes, the cell-by-cell details are not. The precise specification and wiring of the brain depends upon factors not encoded in the genes (called epigenetic factors), including the effects of the environment. In this case, as we will discover, the word "environment" is used broadly to encompass everything from the chemical environment of the womb to sensory experience starting in the womb and continuing through childhood as the brain matures.

The central issue here, the relative contribution of genetic and epigenetic factors to brain development, may sound esoteric, but it is at the core of a debate which has been raging since before Darwin's time: the famous and often bitter debate about whether "nature" or "nurture" is more important in the determination of human mental functions and personality. Over the last 150 years or so, the pendulum of scientific thought has swung at various times to both extremes. Some extreme nurturists, such as the founder of behaviorist psychology, B. F. Skinner, have claimed that the human brain is a "blank slate" with no genetic constraints and that human cognition and personality are entirely formed by experience, particularly early experience. On the other side of the debate have been the extreme naturists (not to be confused with people who like to go bungee-jumping in their birthday suits), a group that has included such historic figures as William James. Naturists have claimed that human mental traits and personality are largely determined by genes, and that barring extreme environmental events, such as being locked in a dark room for long periods, early experience does not significantly contribute.

The debate continues today, but the range of views has trended toward the middle. Now fewer scientists inhabit either extreme pole of the nature-nurture spectrum. In part, this stance has come from accumulating evidence that, for some mental and behavioral traits, there is a clear contribution of genes. A portion of this evidence has come from studies of genetically identical twins (called monozygotic twins by biologists) separated soon after birth and raised by different families. For example, identical twins given psychological tests to pinpoint personality traits, such as extroversion or conscientiousness or openness, showed that identical twins have tended to share many of these traits whether or not the twins were raised together. These studies have been performed by now in a number of different countries, mostly in the more affluent parts of the world.

Not surprisingly, tests of "general intelligence" in adopted twins have generated a lot of controversy. Early studies on this topic were sloppily designed and some even involved scientific fraud. More recently, however, large, carefully designed trials seem to converge on a similar conclusion: in children and young adults from middle-class or affluent families, in studies that have used a combination of twins, identical and nonidentical, raised together and apart, about 50 percent of "general intelligence" can be attributed to genes, with the remainder determined by environmental factors. In other words, genes influence general intelligence but to a lesser degree than they influence personality.

Some telling details emerge from intelligence tests of particular twin subgroups. For example, when identical twins are adopted into different families and one of those families is extremely poor, the poor twin is much more likely to score lower on intelligence tests. Twins raised in poverty perform worse on intelligence tests than twins raised in middle-class households. But twins raised in middle-class households do not perform worse than twins raised in wealthy households. In other words, for the case of "general intelligence," both genes and environment contribute, but in the extreme case of environmental deprivation seen in the poorest households, the effects of environment become much greater and largely overcome the effects of genes.

In contrast, other behavioral traits do not appear to be strongly influenced by genes: food preferences (in both rodents and humans) are largely determined by early experience and are therefore not similar in identical twins raised apart. Sense of humor is another. Identical twins raised apart tend not to find the same things humorous, whereas they do share a sense of humor with their adoptive siblings. These examples show that blanket generalizations about the contribution of genes to mental traits are not warranted. We must consider different aspects of mental function on their own terms.

Separated identical twin studies have been useful for untangling contributions of genes and environment. But they are not perfect. First, shared environmental factors begin in the womb. If, for example, the mother has a high level of stress-induced hormones in her bloodstream during pregnancy, this affects the development of both twins. This is an example of a biological influence that is not genetic, but epigenetic. Second, although the phrase "separated at birth" has become engrained in our popular culture, in practice, such separation rarely occurs. Most twins are adopted after spending days to weeks (and sometimes even months) together, sharing the same nursery environment. Third, some separated twins who have been recruited for these studies have been reunited for some time before their participation in the study. What this means is that direct comparisons between identical twins raised apart and identical twins raised together may overestimate the contribution of genetic factors. However, comparisons between identical twins raised apart and nonidentical same-sex twins (dizygotic twins) raised apart should not be biased in this way because these factors will apply equally to both groups. Indeed, the tests mentioned earlier have shown that identical twins raised apart are significantly more alike in measures of personality than nonidentical twins raised apart. Thus, at present, it is clear that for some human behavioral traits, there is a significant contribution of genes.

Another major factor that has moved many scientists closer to common ground in the nature-nurture wars has been a better understanding of how genes and environment interact in brain cells. In the past, there has been a tendency to imagine that genes and behavior interact in only one direction: genes influence behavior. We now know that the environment, broadly considered, can also influence gene function in brain cells. In other words, nurture can influence nature and vice versa. Causality, in the brain, is a two-way street.

Let's briefly review a little molecular genetics to help understand how the environment can influence genes. Each cell in the human body contains the complete human genome, all 23,000 or so genes, arranged on strands of DNA organized into 23 chromosome pairs (one set from Mom and the other from Dad) in the cell nucleus. Each gene consists of a series of DNA bases that provide the information that ultimately directs the construction of a chain of amino acids. These chains of amino acids are called proteins. Proteins form the important structural and functional units of the cell. For example, they make all of the important neuronal molecules discussed so far. These include ion channels (such as the voltage-sensitive sodium channels that underlie the upstroke of the spike), enzymes that direct chemical reactions to produce or break down neurotransmitters (like the enzyme acetylcholinesterase, which breaks down the neurotransmitter acetylcholine), and neurotransmitter receptors (such as glutamate receptors), as well as the structural molecules, the cables, tubes, and rods of protein that give neurons their shape.

Every cell in your body has, encoded in its DNA, the information to make all proteins encoded in the genome. But, at any given time, a particular cell in your body is only actively making proteins from a small subset of these genes. A small number of genes make products that are continually needed in all cells in the body. These "housekeeping genes" are always on, directing the production of their proteins. Other genes are activated only in certain cell types. For example, the cells that line your stomach are not producing the proteins needed to grow hair, and your hair follicles are not producing the proteins involved in the secretion of stomach acids. Still other genes may be switched on or off at certain points in development or in response to particular signals, and these are the ones in which we are most interested.

Gene expression is the process by which genes are turned on and off. The molecular mechanisms that underlie it are complex and represent an entire subfield of biology. In brief, however, one or more sequences of DNA called

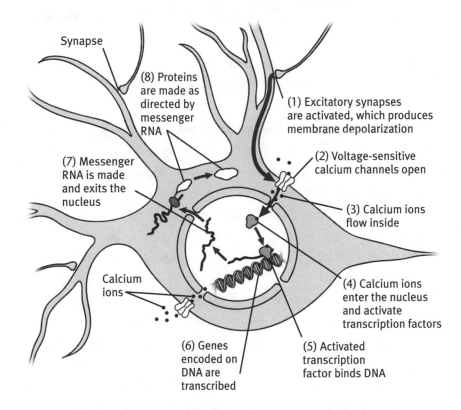

Synapse

(8) Proteins are made as directed by messenger RNA

(7) Messenger RNA is made and exits the nucleus

(1) Excitatory synapses are activated, which produces membrane depolarization

(2) Voltage-sensitive calcium channels open

(3) Calcium ions flow inside

Calcium ions

(4) Calcium ions enter the nucleus and activate transcription factors

(6) Genes encoded on DNA are transcribed

(5) Activated transcription factor binds DNA

FIGURE 3.2. The molecular basis of the link between nature and nurture in the brain. Experiences activate sensory systems, which cause neurons to fire excitatory synapses. This causes a brief increase in the concentration of calcium ions that, through an intermediate biochemical process, activates certain transcription factors, causing them to bind to promoter regions of certain genes and activate them. When the gene is activated it then produces messenger RNA that instructs the synthesis of proteins, the final step in the gene-expression cascade. *Joan M. K. Tycko, illustrator.*

promoters must come into play in a section of DNA adjacent to the region of the actual information-containing part of these genes. Promoters are activated by a set of molecules called transcription factors. Typically, a given promoter will have a specific transcription factor that binds to it. Sometimes, in order to activate a given gene and start the series of events that will ultimately result in making its encoded protein, a particular set of transcription factors must all bind and activate their respective promoters at the same time.

Transcription factors can be activated in different ways. For example, when a rat has been kept in the same cage for weeks and is then placed in a new cage with different sights and smells, a set of neurons in the cortex and the hippocampus will fire bursts of spikes in response to the novel environment. When these neurons fire spike bursts, they cause the opening of voltage-sensitive calcium channels and calcium rushes into the cell. The increase in internal calcium concentration can activate a set of biochemical signals that ultimately result in the activation of transcription factors, one of which is called SRF. SRF binds to a promoter present in many different genes called an SRE. Activation of the SRE promoter is usually not sufficient to activate a gene by itself, but it can be one of several required events. Some other transcription factors are molecules that come from outside the cell, penetrate the outer membrane, and get into the cell nucleus to directly bind promoters. Many hormones, such as the female sex hormone estrogen or thyroid hormone, work in this way.

So, transcription factors acting on promoters provide a biochemical mechanism by which experience, in all of its forms, can affect genes, not by altering the structure of the genetic information, but by controlling the timing of gene expression. It should be mentioned that while transcription factors are one important way of controlling gene expression, they are not the only way. There are several additional steps between switching a gene on and the production of protein, and each of these steps is also subject to regulation. I won't go into all of the ways this can happen, but the larger point here is that many biochemical pathways enable experience to influence gene expression.

HAVING SET THE stage through a consideration of some points in the nature-nurture debate, let's now follow the brain through its development, first in the womb and then during early life. The fertilized ovum begins dividing to form a ball of cells that implant in the lining of the womb several days later and ultimately flatten to form the embryonic disk, a thin pancake-like structure about 1 millimeter in diameter. The ectoderm is the surface layer of cells in the embryonic disk. Over the next few days a portion of the ectoderm receives chemical signals from surrounding tissue that cause it to form the neural plate, a structure in the center of the disk. As the whole embryo grows, the edges of the neural plate curl up and fuse together to form a tube. This neural tube will ultimately become the brain at one end and the spinal cord at the other. The hollow core of the neural tube eventually forms the ventricles, fluid-filled spaces in the center of the brain and spinal cord. This is the state of affairs at about 1 month after conception.

At this point, the neural tube is composed not of actual neurons, but rather of about 125,000 so-called neuronal precursor cells. These cells divide repeatedly, at a furious pace, and give rise to yet more neuronal precursor cells. The rate of cell division in the developing human nervous system is staggering, with about 250,000 new cells being created per minute throughout the first half of gestation. Most of this cell division is happening deep in the developing brain, adjacent to the fluid-filled ventricles. A precursor cell may have several fates. It may divide again to make more precursors, it may become a neuron, or it may become a glial cell. The factors that determine precursor cell fate are critical to determining the ultimate size of the brain and the relative size of its regions.

That brain size in humans is strongly influenced by genes has been known for many years. More recently, the use of sophisticated brain scanners has not only improved the accuracy of brain size measurements, but also allowed scientists to separately measure those parts of the brain composed mostly of bundles of axons (called white matter) versus those parts of the brain composed mostly of neuronal cell bodies and dendrites (gray matter). Impressively, identical twins, whether raised together or apart, are 95 percent similar in gray matter volume. Nonidentical twins, who share the same degree of genetic similarity as any two siblings, are about 50 percent alike in this measure.

This strong finding suggests an obvious question: Can we identify particular genes that control the number of divisions of precursor cells during brain development and thereby influence brain size? In recent years, a small number of candidate genes have emerged in this search. The function of these genes has come to light through investigations of human populations in which a rare and incurable disorder called microcephaly has been found to run in families. Microcephaly is a severe genetic disease that results in a brain that is only about 30 percent of normal size. It does not just represent the low end of the normal range of brain sizes. Rather, adult microcephalics typically have brains about the size of the brain of a chimpanzee or, suggestively, the size of the brain of our 2.5-million-year-old hominid ancestor, *Australopithecus africanus.*

Analysis of microcephalics has revealed a handful of genes that harbor mutations linked to this disease. Of these, we presently know the most about one. The ASPM gene produces a protein involved in cell division: in particular, it helps to form a structure called the mitotic spindle, essential in dividing cells so that each new cell gets its proper share of chromosomes. An important part of this protein is a segment that binds a messenger molecule called calmodulin. The calmodulin-binding region is present in two copies in the ASPM gene of the roundworm, 24 copies in the fruit fly, and 74 copies in humans. Furthermore, careful base-by-base analysis of the ASPM gene in humans, chimpanzees, gorillas, orangutans, and macaque monkeys has suggested that evolution of the ASPM gene, particularly its calmodulin-binding region, has been particularly accelerated in the great ape family. The greatest degree of selective change in the ASPM gene is found along those ape lineages leading to humans. Thus it is likely that the ASPM gene and similar genes have played central roles in the evolutionary expansion of human brain size. You can bet that in the near future brain researchers will look carefully at variation in the ASPM gene and related genes to see if this predicts variation in brain size over the normal range.

As the brain develops during gestation, it is not merely accumulating a larger and larger disorganized mass of cells, but is also initiating important changes in brain shape and the emergence of particular regions (see Figure 3.3). By the end of the second month of pregnancy the neural tube has developed three swellings. The front one will ultimately expand to form the massive and infolded cortex (and some other nearby structures). The lower portions of the neural tube will develop two sharp right-angle bends resulting from different regions adding cells at different rates, and these bends will help to pack the lower portions of the brain into their appropriate orientations. Certain regions will sprout outgrowths that can become quite large, such as the cerebellum, which extends from the back of the developing brain. At the time of birth, many of the neurons that constitute the adult brain have been created. But the brain at birth is far from mature because much of the fine wiring is still to come.

The swelling and bending of the neural tube to delineate regions such as the cortex, midbrain, and cerebellum are under the control of a set of "homeotic" genes that are master regulators of early development. Homeotic genes code for proteins, and these proteins are, you guessed it, transcription factors. Because these transcription factors can contribute to the activation of many other target genes, including those that form the boundaries between regions and that cause groups of cells to clump together, homeotic genes can have widespread effects.

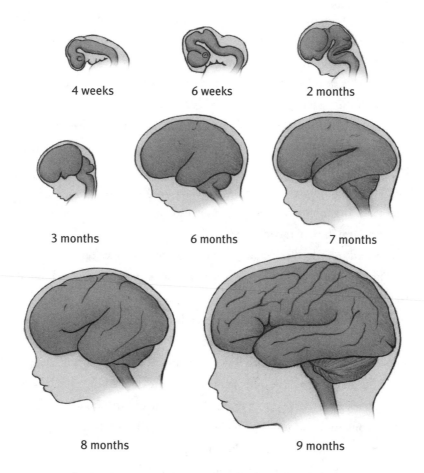

4 weeks   6 weeks   2 months

3 months   6 months   7 months

8 months   9 months

FIGURE 3.3. The development of the brain from 4 weeks after conception, when the neural tube has just formed, through birth. Intermediate stages show the formation of swellings in the neural tube and the expansion and bending of the tube that ultimately give rise to the brain of the newborn. The drawings of the earliest stages are magnified relative to those of the latest stages: the 4-week-old neural tube, for example, is only about 3 millimeters long. Adapted from W. M. Cowan, The development of the brain, *Scientific American* 241:113–133 (1979). *Joan M. K. Tycko, illustrator.*

Brain Surface

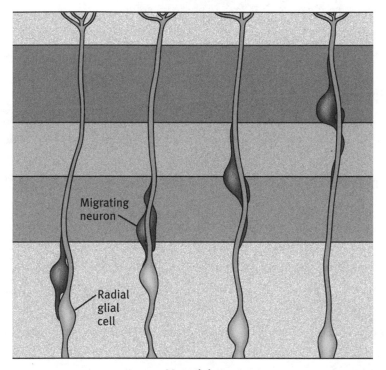

FIGURE 3.4. The migration of newly created neurons. They crawl along radial glia to reach their appropriate locations in the cortex. The radial glial cell functions as a scaffold that stretches from the fluid-filled ventricle, where neuronal precursor cells divide, all the way to the surface of the brain. There, a migrating neuron comes to reside in a layer of the cortex close to the brain surface. Interestingly, in the cerebellum, this process is reversed. Newly born neurons migrate along the outside surface of the cerebellum and then begin crawling along radial glia to migrate inward. Adapted with the permission of Elsevier from A. R. Kriegstein and S. C. Noctor, Patterns of neuronal migration in the embryonic cortex, *Trends in Neuroscience* 27:392–399 (2004). *Joan M. K. Tycko, illustrator.*

Interfering with the action of homeotic genes through mutations or drugs will cause massive and often fatal flaws in brain development.

Once neural precursor cells are done dividing they must migrate from a specialized region for cell division (which is next to the ventricles) to their final location in the brain. The molecular cues that guide neuronal migration are not completely understood, but they include adhesive molecules that guide migrating cells and other molecules that repel them. In those regions of the brain that are organized into distinct cellular layers, such as the cerebellum or the cortex, neurons literally crawl along scaffolds formed by a special class of glial cell, radial glia that extend from the ventricles to the brain surface (see Figure 3.4). The layers are generated in the following way: those cells created first migrate a short distance to the nearest part in the developing cortex while those neurons born later will crawl through the earlier cells to wind up nearer the cortical surface. In this way, the cortex develops in an inside-out fashion with the first neurons created residing in the deepest cortical layers. This complex process can go awry. The effects of errors in migration are less severe than those of defective homeotic genes, but are still very serious: aberrant neuronal migration can result in cerebral palsy, mental retardation, and epilepsy.

As the embryo develops, dividing precursor cells of the neural tube ultimately must give rise to all the diverse types of neurons in the brain. Neuronal diversity encompasses a wide range of characteristics including shape, location, electrical properties, and the neurotransmitter(s) to be used. Somewhat later, of course, the neurons with all of these characteristics have to be appropriately wired together by extending axons and dendrites. Right now, let's focus on the determination of these earlier neuronal properties. One could imagine a plan in which newly created neurons are not restricted to any fate at all. In this view, neurons are generic and multipotent: the properties of individual neurons are determined entirely by their ultimate location in the brain and the signals they receive from surrounding cells. Alternatively, neuronal precursors could become divided into lineages such that, after a certain number of divisions, all of the daughter cells of a particular precursor (and their daughters too) will only give rise to one type of neuron.

To put this into a real context, let's turn to the cortex. Deep in the cortex are a class of neurons, called layer 5 pyramidal neurons, that look kind of like carrots with the narrow end pointed up. Layer 5 pyramidal neurons have a long thick main dendrite and smaller branching dendrites that tend to point up or down but not to the side. These neurons use the neurotransmitter glutamate and receive synapses from the thalamus. Closer to the surface are a different set of neurons, layer 2 cells. When early neuronal precursor cells from a rat that would normally have become layer 5 cells are labeled with a bright green dye to track them and are then transplanted into layer 2 of another rat's cortex, they adopt the properties of layer 2 cells. This result supports the former model, in which developing neurons are derived from multipotent progenitors. But when the experiment is done in reverse and later precursor cells that would normally become layer 2 cells are transplanted into layer 5, those cells do not settle into position and grow to become layer 5 cells. Rather, they migrate out of layer 5 to find layer 2 and grow there in the appropriate fashion. This finding supports the latter model, in which neuronal fate is determined by cell lineage. Although these examples are taken from the cortex, this general theme also applies to other regions of the brain: a combination of local signals and cell lineage factors controls the generation of neuronal diversity. It turns out that, in the end, the story is quite complex: the relative contribution of these factors varies by brain region, cell type, and stage of development.

To this point in our discussion of brain development we have talked a lot about the influence of genes and not at all about the influence of environment. There is a reason for this. Early in development genes direct most decisions about the formation of the brain. Opportunities for environmental influence gradually increase as development progresses, both in the womb and postnatally. In contrasting the role of environment in early versus late brain development, it is useful to distinguish between permissive and instructive influences. The early fetus has no sensory apparatus to carry messages from the world outside, and is entirely dependent upon the maternal blood supply for energy, oxygen, and the molecular building blocks for making new cells. These are permissive factors: interruption by, say, a poor diet, placental malfunction, or maternal disease, can be devastating to fetal brain development. But if these basic fetal needs are met, no information is imparted by these factors that can specifically guide or instruct brain development.

Another form of environmental influence on brain development is through circulating hormones. If the mother is under stress for any reason, from social factors (job loss, a death in the family) to infection, stress-induced hormones will pass into the fetal circulation, where they can influence neurogenesis and migration. The immune system of the mother may also influence brain development, not only through the production of antibodies, but also through a set of molecules, cytokines, produced by the mother's immune system but that can bind to cytokine receptors on fetal neurons. Things get even more complicated when we consider twin fetuses. Hormones produced by one twin can affect the brain development of the other.

Early brain development can also be massively influenced by maternal use of some drugs (both therapeutic and recreational) and alcohol and can be more subtly affected by nicotine. Interestingly, not all of the drugs that can influence fetal brain development are drugs taken to affect the mother's brain function.

For example, certain antibiotics and even acne treatments can have significant effects on fetal brain development.

IN THE LATER stages of pregnancy, while creation of new brain cells continues along with migration and specification of neuronal type, the really hard problem emerges: how to wire the neurons together properly. Here's the difficulty: Not only must neurons from, say, the eye, project to the appropriate visual part of the brain (a particular region of the thalamus that in turn sends axons to the visual part of the cortex in the far back portion of the brain), but also the spatial relationship between adjacent points on the retina, where light is sensed, must be preserved as axons from the eye go into the brain. Otherwise, the visual world would be all scrambled and it would not be possible to construct a visual image of the outside world. This is not just a problem for the visual system. Other sensory systems also have orderly representations of sensory information that must be preserved as brain regions wire together.

Classic experiments published in the 1940s revealed some important aspects of how the brain wiring proceeds. Roger Sperry of Caltech rotated one eye in developing frogs 180 degrees in its socket, before the axons from the eye began to grow into the brain (Figure 3.5). What he found was that even in the case of a rotated eye, the axons from the eye solved the problem of finding their normal targets in the brain's visual center. For frogs this is the optic tectum, which is the equivalent of our human visual midbrain, discussed in Chapter 1. It seemed as if each neuron in the eye was able to find its appropriate target in the optic tectum using chemical cues even when eye rotation disrupted physical cues. Sperry concluded that there are synapse-specific chemical nametags that match axons to their target dendrites of cell bodies during development.

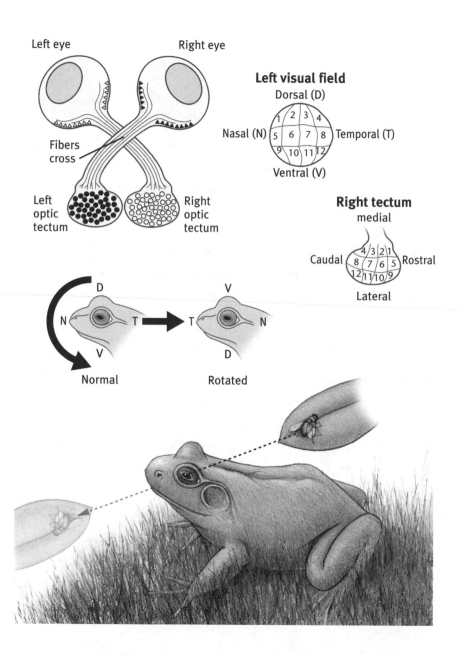

Left eye

Right eye

Left visual field

Dorsal (D)

Nasal (N)

Temporal (T)

Ventral (V)

Fibers cross

Left optic tectum

Right optic tectum

Right tectum

medial

Caudal

Rostral

Lateral

D

N

T

V

Normal

V

T

N

D

Rotated

The general idea that chemical cues can guide appropriate synapse formation has held up well under scrutiny. But the evidence for precise, individual synaptic nametags is poor. For example, if instead of rotating the eye, an experimenter destroys half of the optic tectum in the frog's brain, then all of the ingrowing axons from the eye will crowd into the remaining part of the tectum. This is inconsistent with a nametag model (which would predict that half of the axons would not find their predetermined targets), and suggests an alternative in which gradients of molecules expressed on the surface of neurons in the target region guide the ingrowing axons. Indeed, in recent years some of the molecules establishing these gradients have been discovered, and it has been shown that perturbation of these molecules can disrupt orderly synapse formation. It turns out that a number of the molecules that guide axon outgrowth, by attracting and repelling the tips of growing axons, are the same molecules that guide migration of neurons slightly earlier in development.

But can gradients of guidance molecules determined by genes completely solve the problem of brain wiring? The answer is no. Although early in develop-

---

FIGURE 3.5. The wiring of the visual system in a frog. The top left panel shows a top-down view of the visual system of a frog with the eyes at the top and the visual part of the brain, the optic tectum, at the bottom. The neurons from the retina cross when projecting to the brain so that the right eye goes to the left brain and vice versa. More important, as shown at the top right, the retina is mapped onto the optic tectum in a precise way that preserves the integrity of the image of the visual world formed on the retina (although it is flipped right to left). When Roger Sperry rotated the eye of a frog 180 degrees and then let the axons grow into the brain, the axons from the retina still found their appropriate targets in the tectum. As a consequence, the frog now has an inverted map of its visual world and it will strike in the wrong location when trying to catch a fly for lunch. Adapted from John E. Dowling, *Neurons and Networks,* 2nd ed. (Belknap Press, Cambridge, 2001). *Joan M. K. Tycko, illustrator.*

ment the sensory organs were not yet functional, later, during the time of wiring, the sensory organs are starting to work and the brain itself is becoming increasingly electrically active. Some senses, such as hearing and touch, are quite functional in the later stages of human pregnancy. In the case of fetal vision, there may not be much to see in utero, but there is evidence that even in the absence of light there are spontaneous patterns of activity that sweep across the retina in waves. Electrical activity that results from these spontaneous waves is then conveyed by the developing axons to cause transmitter release in the brain's vision centers.

So, what role does neuronal activity play in wiring up the brain? Let's examine two key observations that will help address this question. First, we'll consider a mutant mouse created in the laboratory of Thomas Südhof at the University of Texas Southwestern Medical Center. This mouse lacks a protein in presynaptic terminals that is essential for the fusion of synaptic vesicles with the presynaptic membrane. As a result, this mouse completely lacks neurotransmitter release and so activity in neurons cannot be propagated to their neighbors. If neuronal activity were essential in the basic wiring of the brain, one might imagine that the brain of this mutant mouse would be a complete mess, with axons and dendrites running every which way. It turns out that, ultimately, this mouse *is* a disaster: it dies at birth because it cannot control the muscles used for breathing. But when its brain was examined at birth and shortly before, there was a big surprise. The wiring plan of this mouse's brain develops in a basically normal fashion. Axons generally project to the right places, and in layered structures such as the cortex neurons are properly arranged and synapses are formed, although in somewhat fewer numbers than normal. Although the brain looks basically normal up to the point of synapse formation, in the days following synapse formation, there is massive neuronal cell death. It is as if, in the absence of receiving synaptic transmission, many neurons could not continue to live. This finding strongly suggests that for much of the brain, initial wiring can occur without neuronal activity.

The second observation concerns the wiring of the brain in adult humans who have been deaf from birth owing to a genetic defect in the cells of the inner ear. In these people, both brain imaging and postmortem anatomical studies reveal that axons from neurons in the visual part of the thalamus that would normally be confined to the visual part of the cortex (located in the far back of the brain) are also found in the auditory cortex (on the sides of the brain). In normal development, a few axons from the visual thalamus stray into the auditory cortex early on, but they are eliminated over time. In congenitally deaf people, visual axons are not only retained, but they sprout new branches. It is as if lack of auditory activity in the auditory cortex allows axons from the visual thalamus to invade new territory and make synaptic connections. This probably happens in a competitive fashion because the unused auditory axons gradually wither from disuse.

These two examples are representative of a large number of similar findings leading to the general conclusion that in most brain regions, large-scale wiring (getting the right axons to the right brain region) and gross maps (getting the axons to the right sub-area of the brain region) are genetically specified. Genetic specification does not involve individual molecular nametags for synapses that would, for example, instruct retinal neuron #345,721 to make a synapse with visual thalamus cell #98,313. Rather, gradients of axonal guidance cues are present that convey more general information to the ingrowing axons. In contrast, the fine details of wiring (getting the axon to make particular synapses with particular individual neurons) is the stage in which experience, as encoded by neuronal activity, plays a role. Genetically determined large-scale aspects of wiring neurons generally occurs earlier in development, while environmentally determined fine details of brain wiring occur later. In the case of humans, the period when brain wiring affects fine-scale brain development starts in the later stages of pregnancy and continues through the first few years of life.

UP TO NOW, I have largely ignored the admittedly important event of birth, straying on both sides of the natal line in discussing brain wiring. In large part this is warranted, because there is no evidence to date for a dramatic or qualitative difference in human brain development that accompanies birth. Rather, the maturational processes of late pregnancy continue on a similar trajectory in newborns. The most important thing about birth from the point of view of brain development is a straightforward consideration: the baby's head has to get through the birth canal and this limits the size of the brain at birth.

Here the inefficiency of brain design becomes painfully apparent to the birthing mother. The reason that Mom has to struggle to squeeze out that big head is directly attributable to suboptimal brain design: the human brain has never been redesigned from the ground up and is therefore spatially inefficient (for example, it has two visual systems, one ancient and one modern, as discussed in Chapter 1), and because it is built out of neurons that are slow, inefficient processors (Chapter 2), a human needs to employ massive interconnected network processing using about 100 billion neurons and 500 trillion synapses. Hence, a big head.

At birth, the volume of the human brain is about 400 cubic centimeters, or about the size of an adult chimpanzee's. It will continue to grow quite rapidly until about the age of 5, at which point the brain reaches about 90 percent of its maximal size. After the age of 5, the brain continues to grow at a slower rate until stabilizing at about the age of 20. The period from birth to the age of 20, in which the brain is increasing in size by more than 300 percent, is accompanied by a host of changes in brain structure. A subset of glial cells in the brain are secreting myelin, an insulating substance that wraps around axons to accelerate

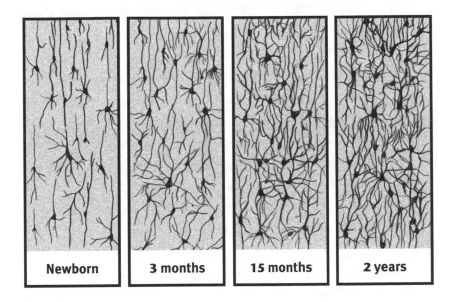

| Newborn | 3 months | 15 months | 2 years |

FIGURE 3.6. Maturation of the human cortex in early life. Although the number of
neurons changes only slightly, the axons and dendrites of these neurons
become much more elaborate. This drawing shows a representative subset
of neurons. It also omits glial cells, which, if shown, would fill in most of
the space between the neurons. Adapted from J. L. Conel, *The Post-natal
Development of the Human Cerebral Cortex,* vol. 1 (Harvard University
Press, Cambridge, 1939). *Joan M. K. Tycko, illustrator.*

spike propagation and reduce energy usage. Myelin secretion causes an increase
in the volume of the white matter. In addition, this is a period of extensive
branching and elaboration of dendrites and axons (Figure 3.6) accompanied by
the formation of many, many new synapses.

In general, the increase in brain volume after birth is not accompanied by an
increase in the number of neurons. A small fraction of the brain's complete
population of neurons is newly created in the first year of life, but some neurons
die off in this period, leaving the total number basically unchanged. If we count

the total number of neurons created during brain development, both before and after birth, we find that about twice as many neurons are created as ultimately reside in the mature brain.

What happens to these extra 100 billion neurons, most of which die before birth? The answer reveals a lot about how electrical activity contributes to the fine structure of brain wiring. Basically, the developing brain is a battleground. There is a competition for survival among neurons that is well encapsulated in the popular phrase "Use it or lose it." What this means is that during development more neurons are created than can actually be used, and, in general, the ones that survive are the ones that are electrically active. The way a neuron becomes electrically active is by receiving synapses that release neurotransmitters and thereby cause it to fire spikes. So, if we look in a bit more detail, the battle is being fought not at the level of whole neurons, but on a smaller scale at the level of synapses. Recall that synapses that are not used tend to wither away (like the synapses conveying auditory information in deaf people), while synapses that remain active are maintained. This encompasses a portion of the idea of synaptic competition, but it is not everything. A synapse can "lose" and be eliminated even if it is active to some degree, if its neighbor is much more active. Strong activation of a synapse not only preserves and strengthens it, but also makes its neighbors weaker and ultimately can cause them to be eliminated. I'll talk a lot about the molecular basis of how this happens in Chapter 5, when I consider memory storage that reuses these same mechanisms.

Can we then envision environmental molding of brain development as a process by which experience selects from a preexisting set of synapses and neurons, keeping some active ones and killing off some (relatively or absolutely) inactive ones? Does the sculptor of experience chisel away at the block of stone that is the developing brain to create the mature form? This is an idea that has been very attractive to certain brain researchers, computer scientists, and even some philosophers. It has been given names such as "selectionist theory" or "neural Darwinism." Although at some level these ideas are correct, they are far from complete. There is now excellent evidence from different animals, brain regions, and conditions that experience-driven electrical activation can cause axons to sprout new branches that will develop new presynaptic terminals. This can also occur on the postsynaptic side: electrical activity can cause the formation of new dendritic spines and small dendritic branches. So, if the brain is a block of clay, then experience sculpts it not just by carving away inactive or ineffective parts but also by sticking on new bits in the form of newly created wiring (axons, dendrites, and synapses) in active regions.

The ability of the brain to be modulated by experience is called neural plasticity. The degree of neuronal plasticity will vary depending upon the brain region and the stage of development. This gives rise to the idea that there are critical periods during which experience is necessary to properly sculpt neural circuits for certain brain functions. One of the best examples comes from vision. If a baby has an eye closed with a bandage (to treat an infection, for example) and the bandage stays on for a long time, then that baby can be blinded in that eye for life. The same bandage applied to the eye of an adult will cause no lasting problem. The reason for the blindness is not that the eye has ceased to function (this can be confirmed by recording light-evoked electrical activity from the eyeball) but rather that the information from that eye was not present to help retain the appropriate connections in the brain during the critical period for vision.

There are other forms of neural plasticity that are not subject to a clearly delimited critical period. In the early 1960s neural plasticity was not a widely considered topic. Most scientists thought that the brain had a set of connections which were wired like a circuit board in a radio and not subject to change. So it was quite a shock to the scientific community when Marion Diamond and her

Deprived **Enriched**

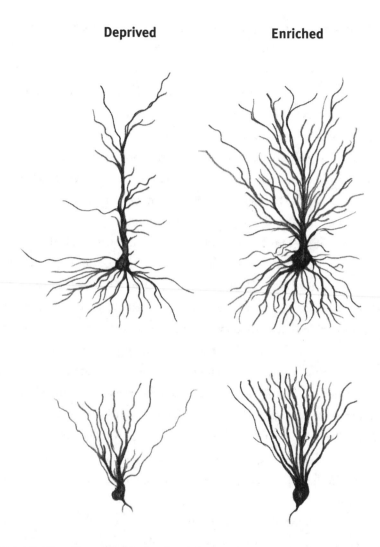

FIGURE 3.7. The effect of deprived and enriched environments. Deprived environments can reduce the dendritic complexity of neurons in the cortex and hippocampus. Adapted from C. J. Faherty, D. Kerley, and R. J. Smeyne, A Golgi-Cox morphological analysis of neuronal changes induced by environmental enrichment, *Developmental Brain Research* 141:55–61 (2003). *Joan M. K. Tycko, illustrator.*

coworkers at the University of California at Berkeley analyzed the brains of adult rats that had been removed from their boring, prison-like, individual lab cages and housed in an "enriched environment" with toys, places to explore, and other rats. After several weeks in the enriched environment, these rats were killed and their brains prepared for microscopic examination. In several cortical regions, neuronal dendrites were larger and much more highly branched and there were more dendritic spines and more synapses than in rats kept in Spartan conditions. This suggested that even the adult brain was much more plastic than anyone had imagined at that time.

Crucially, this process was reversible. Rats placed in the enriched environment for several weeks and then returned to standard lab cages for several more weeks had cortical neurons that looked like those of rats that had never left the standard cages. It is tempting to leap to the conclusion that a similar strategy of "environmental enrichment" would be beneficial to children. The thing to keep in mind here is that the so-called enriched environment for rats is actually just a simulation of what rats encounter in the wild. The standard lab cage is deadly boring: for the rat it's like being in solitary confinement. Rather than showing that extra enrichment beyond normal experience can boost brain growth, what this experiment shows is that severe environmental deprivation can, at least temporarily, cause a reduction in the complexity of cortical circuits.

Are there correlates of brain-critical periods that can be reflected in higher cognitive processes? There is evidence to suggest that a critical period exists for language acquisition. Babies under the age of 6 months appear to be able to distinguish all forms of speech sounds from any language. But after 6–12 months of exposure solely to Japanese, infants have begun to tune into the sound differences important for Japanese speech and to ignore other distinctions that are not present in Japanese (such as the "r" versus "l" sounds in English). The flip side of this phenomenon is that babies exposed to two languages can develop perfectly accented speech in both languages.

Children learning a second language after they are 5 years old or so can do very well, but are likely not to have perfect accents. School-age children generally will acquire second or third languages more easily than adults. A few studies suggest that the window for development of a *first* language closes at around 12 years. But these claims depend upon a handful of attempts to teach language to abused children who were essentially locked away for large parts of their childhood. A 12-year-old who had suffered in this way was able to acquire only the most basic rudiments of language, while a 6-year-old subjected to similar deprivation and abuse eventually learned language quite well. The problem is that these children obviously were deeply traumatized in ways that could strongly influence subsequent social learning. Therefore, it is hard to make a straightforward conclusion about language development from their tragic cases.

Is it possible to identify critical periods for other forms of learning, and, if so, can measurements of various brain structures during early life inform this process? At present there is an explosion of interest in so-called brain-based education. References to developmental neurobiology have been used to justify everything from a whole-language approach to reading (as opposed to a phonics-based curriculum) to assessment of a student's work portfolio as a teaching tool. A report from the 1997 White House Conference on Early Brain Development states, "By the age of three, the brains of children are two and a half times more active than the brains of adults—and they stay that way throughout the first decade of life . . . This suggests that young children—particularly infants and toddlers—are biologically primed for learning and that these early years provide a unique window of opportunity or prime time for learning."

Unfortunately, brain research is being invoked here to justify policies that may or may not be valid, but current neurobiological knowledge can add little to the debate. For example, if we wanted to predict a critical period for learning arithmetic, it's not clear where we should look in the brain and what we should look for once we determine a location. Even the general statement from the White House conference is problematic. First, the actual evidence for a 2.5-fold increase in brain activity in normal 3–10-year-olds is almost nil, and, second, even if it were true there is no particular reason to believe this indicates a unique opportunity for learning. One could just as easily imagine that this increased activity represents "background noise" in the brain that might interfere with learning and that such a finding could justify shifting more educational resources to teaching older children. This is what can happen when a tiny bit of science finds it way into a policy debate.

It is clear that experience early in life is important for developing and fine-tuning the circuitry in certain parts of the brain, but this cannot be used to justify the contention that there is a critical early window for many forms of learning. One problem with analyzing learning in early life is that it's hard to distinguish a super-plastic state in early development in which learning might be particularly effective from the "founder effect" of early information. Learning is a process by which new experiences are integrated with previous experiences. Therefore, early experience may be important, not because it is written into neural circuitry more effectively, but rather because it is the basis for subsequent learning.

Similarly, there is little evidence that efforts of some parents to "enrich" the environments of newborns or toddlers with multicolored mobiles or Mozart CDs will result in any measurable consequences for brain wiring, or for general measures of cognitive function. Essentially, in terms of brain wiring, the evidence to date is that a child's early environment is like your need for vitamins: you need a minimum dose, but beyond that, taking extra won't help. That is, exposure to varied spoken language, narrative, music, and the ability to explore, play, and interact socially are all important for youngsters. But beyond these basic experiences that are present in most middle-class homes there is no reason to believe that further "enrichment" confers any benefit to the structure or function of the developing brain.

So, let's break it on down. We've seen that the initial stages of brain development, proliferation of neuronal precursor cells and migration of these cells to their correct positions, are mostly genetically determined. There are environmental influences at these early stages but they are mostly permissive rather than instructive. Environmental influences during early gestation tend to be revealed with problems such as maternal malnutrition or stress. As development progresses to the stage of wiring the brain, there is a mixture of genetic and environmental influences with genes guiding large-scale wiring and neural activity (deriving from both internal and external sources) guiding fine-scale wiring. Patterns of experience-driven neural activity can influence fine-scale wiring both by eliminating some relatively disused synapses and neurons and by promoting new growth of new axons, dendrites, and synapses. In certain brain regions (such as the visual cortex) there are critical periods in early life where experience must be present or the fine-scale wiring will degenerate and never regrow in later life. In other brain regions, experience-driven neuronal plasticity allows for the fine-scale wiring of the brain to be subtly changed throughout life. In Chapter 5, I will explore how these mechanisms, which underlie the influences of nurture on brain development, have been retained in the mature brain and modified to store memories.

How did brain development come to be a two-way interaction between nature and nurture? This situation has been imposed by three main factors. First, our neurons are slow and unreliable processors. Second, our brains have never been redesigned from the ground up and are therefore filled with multiple systems and anachronistic junk. These two factors work together to require our brains to employ a huge number of neurons to achieve sophisticated computation. Third, this number of neurons is so large that it is not possible to genetically specify each and every synaptic connection with a unique chemical nametag. Therefore, because of informational constraints imposed by brain size, fine-scale brain wiring must be driven by experience rather than genes. Although this means that we must spend an unusually long childhood wiring up our brains with experience (much longer than any other animal), the mechanisms of neural plasticity that have emerged to allow this have also given us our memories and ultimately, our individuality. Not a bad deal, really.

Chapter Four

Sensation and Emotion

EVERY DAY WE go through our lives trusting our senses to provide us with the lowdown: a direct, unadorned view of the external world. In particular, we are inclined to believe vision over our other senses. To illustrate this, we need to look no further than the usage of sensory terms in our casual speech:

"I *see* that the President is a liar."

[This means "the truth about the President is revealed to me."]

"I *hear* that the President is a liar."

[This may or may not be true. It warrants further attention.]

"Something doesn't *smell* right about this President."

[I'm suspicious, but it's hard to say exactly why. It warrants further attention.]

Whichever President we're talking about here, the larger point is that we trust our senses and, of our senses, we trust vision the most—think of "eyewitness" testimony in court. What's more, in everyday life we behave with the implicit assumption that our sensory information is "raw data," and, if necessary, we can evaluate this data dispassionately and, only then, make decisions and plan actions based upon it.

What I hope to convey in this chapter is that this feeling that we have about our senses, that they are trustworthy and independent reporters, while overwhelming and pervasive, is simply not true. Our senses are not built to give us an "accurate" picture of the external world at all. Rather, through millions of years of evolutionary tinkering, they have been designed to detect and even exaggerate certain features and aspects of the sensory world and to ignore others. Our brains then blend this whole sensory stew together with emotion to create a seamless ongoing story of experience that makes sense. Our senses are cherry-picking and processing certain aspects of the external world for us to consider. Furthermore, we cannot experience the world in a purely sensory fashion because, in many cases, by the time we are aware of sensory information, it's already been deeply intertwined with emotions and plans for action. Simply put: In the sensory world, our brains are messing with the data.

SO HOW DOES this sensory manipulation come about? To start with, let's consider some general themes in the organization of sensory systems. These systems are typically organized into maps of the external world. In Chapter 3, I talked about how the rough map of the visual world is created in the brain by gradients of axon guidance molecules during early development and is then refined by experience at a later stage. So, if you look at the place in the cortex where visual information first arrives (called the primary visual cortex) you will find a map of the visual world (in this case, the map happens to be upside-down and backward). What this means is that the far right portion of this area will be activated by light coming from the far left of the field of view, and, conversely, the far left of this area will be activated by light coming from the far right of the field of view, with the intermediate areas of the visual cortex filling in the middle. Other senses also have maps. For hearing, the map is for pitch: if you look at the primary auditory cortex you will find that one end is activated by very high tones and the other by very low tones, with intermediate pitches arranged gradually in between.

These maps, though organized, often reflect the particular anatomy of sensory systems. For example, your retina devotes an unusually large number of light-sensing neurons to evaluating the very center of your field of view (this is why your visual acuity and color vision are better in the center than at the periphery). As a consequence, the map of the visual world in your cortex is distorted so that those neurons responding to light in the center of your visual field take up an inordinate amount of cortical space. An even more dramatic example of this is found in your primary somatosensory cortex, where information about touch and body position first arrives in higher brain centers. We have very good tactile discrimination in our fingers and face, particularly the lips and tongue (hence the popularity of kissing), and relatively poor tactile discrimination in some other locations such as the lower back. This is reflected in the size and arrangement of body parts in the cortical map of the body surface, which is called the sensory homunculus (Figure 4.1, left). ("Homunculus" just means little man.) When the size of the representation in the cortex is used to scale the body parts in a drawing of an assembled homunculus, as seen in Figure 4.1, right, the exaggeration of certain body parts in the sensory map which have fine tactile discrimination becomes even more apparent.

Most people who look at the sensory homunculus long enough will eventually stammer out something like, "Given how sensitive the genitals are, shouldn't

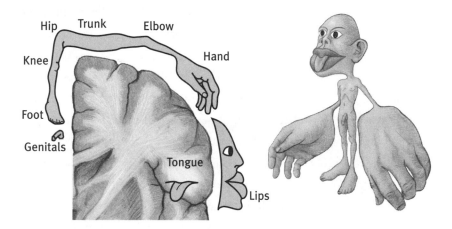

FIGURE 4.1. The representation of tactile sense in the brain. Those body parts that have fine tactile sensation, such as the hand, lips, and tongue, occupy a disproportionate amount of space in the primary somatosensory cortex. Left: A view of the right half of the brain, cut in the coronal plane (from ear to ear) and opened to face us. The body parts that are represented in the cortical map are shown adjacent to or overlying the corresponding regions of the cortex. Note that the map is fractured: some adjacent body parts, such as the forehead and hand, are not adjacent on the body. Right: A view of the human male in which the sizes of the body parts have been scaled to the size of their representation in the primary somatosensory cortex. To me, this little guy looks a bit like Mick Jagger. *Joan M. K. Tycko, illustrator.*

they be larger?" We know that the genitals are sensitive to touch and that there are particular nerves which carry sensory information from the genitals into the spinal cord and up to the brain. One potential explanation for the size issue hinges on the need to be more precise when we say "sensitive to touch." The parts of the homunculus that have huge representations (such as hands, lips, and tongue) are not merely able to detect faint sensations but can also discriminate the location of these sensations very precisely. You might imag ine that these two abilities always go together, but they do not. The ability to do the finest discrimination, necessary for tactile form perception (as in reading Braille), requires a special type of nerve ending in the skin that is abundant in the fingers, lips, and tongue but almost completely absent in either the penis or the clitoris. The genitals, while they can easily detect faint sensations, cannot accomplish tactile form perception. In the spirit of old-fashioned natural philosophy, you can experiment with this at home. In this way, they are somewhat like the cornea of the eye: quite sensitive to faint sensations such as a grain of sand, but without an ability to precisely locate those sensations. This difference in the exact type of touch sensation is likely to explain why neither the cornea nor the genitals (male or female) are particularly large in the sensory homunculus.

SENSORY SYSTEMS IN the brain typically do not have a single map of their world, but rather many, spread over adjoining regions of the cortex. In many cases, sensory information is divided or duplicated and sent to different subregions of the cortex that are specialized to extract particular forms of information. A good example of this is found in the visual system. The cells that send visual information from the retina into the brain can be divided into two types, P cells (P is for "parvi," which means small) and M cells (M = "magni" = large). Each P cell responds to only a small part of the visual scene, and all are sensitive to color. The M cells, which are important for detecting moving stimuli, are insensitive to color, and integrate information from a larger area.

Although P-cell and M-cell signals travel side by side in axons going from the retina to the thalamus and then in other axons from the thalamus to the primary visual cortex, little of this information mixes in either of these locations. After the primary visual cortex, this information becomes clearly divided, as different sets of axons carry P-cell and M-cell information along different routes

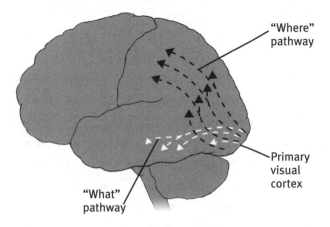

"Where" pathway

Primary visual cortex

"What" pathway

FIGURE 4.2. The processing of visual signals in two different pathways to extract "what" versus "where" information about objects in the world. The figure shows the left surface of a human brain. After a relay in the thalamus, signals from the retina arrive at the primary visual cortex, at the far back end of the brain. The primary visual cortex sends fibers carrying visual information into two different processing streams, each with many individual areas. The higher pathway, into the parietal lobe, is the "where" pathway, specialized to determine the position, depth, and motion of objects. The lower pathway into the temporal lobe constitutes the "what" pathway, which uses visual detail and color to evaluate and identify objects. Adapted with the permission of Elsevier from A. C. Guyton, *Textbook of Medical Physiology,* 8th ed. (W. B. Saunders Company, Philadelphia, 1991). *Joan M. K. Tycko, illustrator.*

(Figure 4.2). Signals from the M cells are conveyed into a set of processing stations in the parietal lobe that are specialized to use this broadly tuned visual information to plot the location, depth, and trajectory of visual objects, both animate and inanimate. This has been called the "where" pathway. P-cell information has another fate. It is sent to a set of regions in the temporal lobe that use finely tuned visual information, including color, to recognize objects, thereby constituting the "what" pathway. In later stages the "what" and "where" streams converge, presumably to allow this information to be integrated in our visual experience.

If we travel the long and winding road of the "what" pathway and examine the responses to visual stimuli at each station, an interesting theme begins to emerge. Neurons in the earliest stations, such as the retina, respond well to simple stimuli, such as dots of light. Somewhat farther along, in the primary visual cortex, the optimal stimuli are more geometrically complex, like a bar with a particular orientation. Further still, along the "what" pathway optimal stimuli are actual real-world objects such as a hand or a rock. It seems as if the visual system gradually builds the ability to detect more complex features and objects through successive processing in the "what" pathway. At the very end of the "what" pathway, information is fed to memory and emotion centers such as the hippocampus and amygdala.

The later stations of the "what" pathway can be quite specialized. Damage to these regions (either from trauma or developmental/genetic problems) can lead to very specific impairments, such as an inability to recognize particular human faces, a syndrome called face-blindness, or prosopagnosia. People who sustain damage to nearby regions can have a failure to recognize visual objects (called visual object agnosia). Milder cases of this can involve inability to recognize a particular object within a class—the inability to pick out one's own car in a full parking lot is typical. More severe cases include the profound confusion of animate and inanimate objects, as in the unfortunate individual who inspired the title of Oliver Sacks's famous book *The Man Who Mistook His Wife for a Hat*. Similar phenomena have been observed in laboratory monkeys that have sustained bilateral damage to the temporal lobe: they may try to eat grossly inappropriate nonfood items (such as a lit cigarette).

So, to put it all together, when you're at a U2 concert and your mother bursts from the wings and runs across the stage in a desperate attempt to smooch Bono, the M-cell driven "where" pathway in your parietal lobes will register that there's something moving on a trajectory that will intercept him, and the P-cell driven "what" pathway in your temporal lobes will recognize this something as Mom. Because the "where" pathway is a bit faster (and has an easier computational job) you will register the former slightly before the latter. Any feelings of embarrassment or elation you may feel in this situation are probably mediated by fibers carrying this information to your amygdala and a handful of other regions involved in emotional responses. Of course, you don't experience this visual scene as separate "what" and "where" information—it's blended into a coherent unitary perception that feels natural and true.

WHAT HAPPENS WHEN information from multiple sensory areas, which is normally kept separate, is blended in the brain? Consider the case of E.S., a 27-year-old professional musician who lives in Switzerland. She is a synesthete, meaning that she has involuntary physical experiences across sensory modalities. In particular, whenever she hears a certain tone interval she experiences a taste on her tongue. This sensation is totally consistent: a major third will always produce a sweet taste, a minor seventh evokes a bitter taste, and a minor sixth, the taste of cream. She also experiences colors in response to tones: C is red, F-sharp is violet, and so on. Careful study by Lutz Jäncke and his co-workers at the University of Zurich has shown that E.S. uses her remarkable synesthetic sense as a memory aid in musical performances.

E.S. is only one of many types of synesthete. Others can hear odors, smell textures, or even feel heat from certain forms of visual stimulation. About half of all synesthetes have more than one cross-modal drive, these are never bi-directional: someone who sees colors in response to particular odors will not also perceive odors in response to colors. By far the most common forms of synesthesia are the perception of color in response to graphemes (written numbers, letters, or symbols) or sounds, particularly musical sounds. Interestingly, though most synesthetes have their cross-modal experience triggered by an explicit sensory experience (say, the number "5" as written in Arabic numerals but not other representations of five such as the Roman "V" or five hash marks "|||||"), some others are triggered by concepts. For example, there is a group of synesthetes who have colors triggered by temporal categories: December is blue while May is red; Saturday is pink and Wednesday is light green.

Synesthetes have normal-to-above-normal general intelligence and they appear typical in personality tests and general neurological exams. They do not hallucinate or show an unusual incidence of mental illness. Determining the number of synesthetes in the general population is difficult, but recent estimates have been as high as 1 in 200 people. Synesthesia is much more common in women and in left-handed people. Although it is hard to exclude sampling bias, it appears, not surprisingly, that synesthetes tend to be drawn to the creative professions, such as writing, visual art, music, and architecture.

Synesthesia has been known for over 200 years, and was even found by Darwin's cousin, the nineteenth-century scientist Francis Galton, to run in families. But until recently, it was a phenomenon that never quite achieved scientific respectability. Many neurologists thought that synesthetics were simply flaky and poetic: they didn't really *experience* cross-modal associations; they just had a flair for metaphoric language. In their view, E.S.'s reporting that the note F-sharp evoked the color violet was not fundamentally different from what a wag said of the poet W. H. Auden, "He's got a face like an unmade bed."

There are several reasons to believe that synesthetes are having genuine cross-modal experience rather than merely making poetic associations. First, the experiences they report do not change over time: they are consistent over many years, even when subjects are tested without warning. Second, some clever perceptual tests have supported the idea of true synesthetic experience. For example, imagine an array of 5's on a piece of paper with a few 2's thrown in, all printed in black type on a white background. If you were asked to count the number of 2's this would require a systematic search and your response time would be slow. If, however, all of the 5's were in red type while the 2's were green, then the 2's would "pop out" in your perception and you could count them much faster. When Edward Hubbard and V. S. Ramachandran of the University of California at San Diego gave the black-type task to number-to-color synesthetes, they solved it rapidly, like normals facing the colored-type task, which supports the notion that synesthetes truly see the numbers as colored. Third, brain imaging studies have shown that synesthetes have activation of brain regions corresponding to their cross-modal sense. Work by Jeffrey Gray and his coworkers at the Institute of Psychiatry in London showed that spoken-word-to-color synesthetes showed activation of both auditory/language regions and centers that process color vision (called V4/V8) in a word-listening task, while normals showed only auditory/language activation.

The suggestion from this and other brain imaging studies is that synesthesia results from the spread of signals from their typical sensory regions in the brain to the regions subserving other senses. The most popular hypothesis to explain how this comes about is that aberrant synaptic connections (say, from the auditory information stream to visual color areas) somehow fail to be eliminated in early postnatal development, and their retention and elaboration in later life drives particular synesthetic experience. This notion is supported by the observation that in the most common forms of synesthesia, such as tone-to-color and grapheme-to-color, we see coactivation of adjacent regions of cortex, while rare forms of synesthesia, such as odor-to-hearing involve coactivation of more distant regions.

Synesthesia is not a disease state. It is likely to represent one end of a spectrum of multi-modal sensory experience: we all integrate sensory information to some degree. Indeed, it is possible that as infants, before the first wave of activity-driven refinement of synaptic connections was complete, we were all once highly synesthetic.

AS WE GO through life, whether attending a concert or walking down the street, we are not generally aware of the convoluted neural architecture of our sensory systems. We just experience the external world and it feels like the truth. In fact, our sensory systems are messing with nearly every aspect of our sensations from their quality to their timing. Millions of years of evolution have biased our sensory systems in very unusual ways. First, we must consider the rather simple fact that the range of stimuli we can detect is merely a subset of possible sensory information. We can see certain wavelengths of light from deep red to deep violet, but not light that is beyond either end of this spectrum. By contrast, many birds can see in the ultraviolet. This allows certain birds of prey (such as hawks) to detect the urine trails left by their prey (field mice or rabbits). Likewise, humans can hear over a certain frequency range (from about 20–20,000 cycles per second), but this range is just one slice of auditory information: bats, whales, and mice can hear much higher tones (up to about 100,000 cycles per second). We can discriminate about 10,000 different odors, but dogs can do much better (250,000 odors is one estimate). The list goes on and on in every sensory modality. Presumably, it has been evolutionarily advantageous for dogs to have this large olfactory range while humans can make do with less information. Our senses are merely "peering through a keyhole" into sensory space.

Evolutionary pressures have influenced not only the boundaries of our senses but also how that sensory information is subsequently processed in the brain. Our sensory systems have adapted in ways that are important for key behaviors such as feeding, avoiding danger, mating, and child care. Although there are many quirks of sensory processing that are unique to particular senses, there are also some general themes. For example, our sensory systems are generally built to give a stronger response to novel stimuli than those which are ongoing, a process called adaptation. You know this from your own experience: If you walk into your kitchen the morning after cooking fish for dinner you may detect a lingering odor initially, but after a minute or so, you barely notice it. If you walk out of the kitchen and then reenter it later, the odor will once again become briefly apparent. Similarly, if you use a computer you might notice the faint high-pitched whine of the hard drive when you first sit down to work, but this is likely to fade from your perceptual world rather quickly. This adaptation is likely to be evolutionarily useful because it allows you to focus on novel, potentially dangerous (or tasty) stimuli out in the world.

Our sensory systems are also specifically designed to detect changes in dimensions other than time. One of the best examples of this is edge enhancement in the visual system, which is useful for distinguishing objects from their background, an understandably adaptive function for finding food or avoiding predation. Edge enhancement is produced by circuitry in both the retina and the subsequent processing stations of the brain that, when assigning a perception of luminance to a given spot in the visual field, makes it appear darker if the surrounding area is brighter. This is produced by a process called lateral inhibition, in which neurons in a visual map inhibit their near neighbors when activated. They do this using axons that form synapses releasing the inhibitory neurotransmitter GABA. Although we are normally unaware of edge enhancement in our visual world, it can be revealed in a number of optical illusions, one of which is shown in Figure 4.3.

Edge enhancement is an example of how our brain distorts our perception of the world to render sensory information more useful. This is all very well, but our sensory systems actually have a much bigger problem: they must make sen-

FIGURE 4.3. An optical illusion produced by circuits in the visual system that are designed to enhance edges. The gray horizontal stripes in the right and left panels are the same uniform shade of grey. The stripe in the left panel, however, appears to alternately become brighter and darker as the visual system integrates information outside of the stripe to enhance edges. To convince yourself that this is really true, cover everything but the horizontal stripe in the left panel and watch it become uniformly gray. *Joan M. K. Tycko, illustrator.*

sory time seem continuous and flowing. At this point, you're probably thinking that time is, by its nature, continuous and flowing, so why does the brain have to do anything to make it seem so? Let me explain what I mean. When you survey a visual scene, your eyes do not stay still. They tend to jump rapidly from point to point. These jumps are called saccades and they function to place different parts of the visual scene in the center of your field of view, where fine form or color discrimination is possible (Figure 4.4). It takes time to complete a saccade: your brain has to issue commands and these commands have to be relayed through several locations to ultimately reach axons that make synapses on your eye muscles and release the neurotransmitter acetylcholine to excite them. Then, the eye muscles have to contract to produce the movement. The longest possible saccade you can do, from the far right of your visual field to the far left,

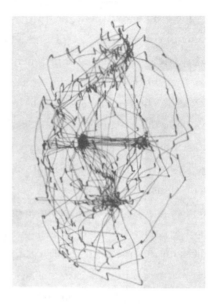

FIGURE 4.4. Saccades involved in scanning a visual scene. An eye-tracking device was used to record eye position as a human subject scanned this photograph of a girl from the Volga region of Russia for about 3 minutes. Most of the line segments show here are saccades (the others are slow tracking movements). From A. L. Yarbus, *Eye Movements and Vision* (Plenum Press, New York, 1967); reprinted with permission from Springer Science & Business Media.

will take about 200 milliseconds (one fifth of a second). During these saccades you do not see the visual world sliding around as your eyes shift, nor does your vision black out during this period. Your retina does not stop sending information to your brain during a saccade. Rather, the signals showing the visual world sliding around during a saccade are sent from your eye to your brain, but they do not make their way into your perception. As you know, your visual perception seems to make sense: it is continuous and flowing and you are generally unaware of your eyes jumping about.

How does your brain fill in the gaps created by saccades to achieve this smooth visual effect? To explain this, it is first necessary to mention that there is a brief delay between the point when events in the world impinge upon the sensory organs (light falls on the retina, sound waves reach the eardrum, odorant molecules bump into sensory cells in the nose) and the point when we become aware of these sensations. There is some variation in this delay, depending on the sense involved and the exact type and intensity of stimulation, but the range is generally between 50 and 300 milliseconds. This delay, like the ones that television networks impose on "live" broadcasts to allow them to bleep out prohibited words, allows the brain to engage in some funny business. It's important to understand that the delay does not simply correspond to the time it takes for the first electrical signals to reach the primary cortex (usually 20–50 milliseconds). In most cases the awareness of sensations requires further cortical processing, and, as a consequence, a bit more time.

So, in the case of a saccade, your brain ignores the visual information conveyed during the eye movement. Then, when the saccade is complete, your brain takes the visual image of the new location and uses it, retroactively, to fill in the preceding time gap. Most of the time you do not notice this at all. But in particular circumstances, this brain trickery can be revealed, as in a phenomenon called the stopped-clock illusion. When you make a large saccade that results in your eyes coming to rest on a clock, it will sometimes appear as if the second hand of the clock then takes slightly longer than normal to move to its next position. For this illusion to work, the movement must be a true saccade (slowly sweeping your eyes across the visual field to ultimately rest on the clock engages a different mechanism in the brain) and the clock must be silent (regular ticking will destroy the illusion). It will work on either a traditional analog clock or a digital clock that shows seconds, but will be most apparent on those trials where the saccade is completed immediately after a clock movement. For a short period, the clock appears to have stopped (a phenomenon called chronostasis) because the brain extends the percept of the new location back in time to just before the start of the saccade.

For many years, it was thought that chronostasis was a strictly visual phenomenon. In the last few years, however, it has been demonstrated for other senses as well. One study that required subjects wearing headphones to shift their attention from their right ear to their left before judging an interval between two tones found a similar phenomenon. This may underlie the so-called dead phone illusion, in which, when a person rapidly shifts attention and simultaneously activates a phone handset, she judges the silent interval before the dial tone to be unusually long. In another experiment it was shown that, following a rapid reaching movement, people overestimate the time their hands have been in contact with a newly touched object. Much as in the visual stopped clock illusion, it appears as if tactile perception were extended backward in time to a moment at the onset of the reach. These findings suggest that chronostasis, resulting from extending perception backward in time to mask a period where perception would otherwise be confusing, is a widespread feature of sensory systems. This is one trick our brains employ to create a useful, coherent sensory narrative.

MANY PHILOSOPHERS AND cognitive scientists approach perception as if it were a completely objective and logical process. In their view, perceptions can sometimes trigger emotions but it is possible to divorce emotion from perception and act on perceptions in a purely unemotional fashion. This perception/emotion distinction resonates throughout our Western cultural tradition. We see this most strongly in medicine, where we have two different fields for treating brain disease. Neurology mostly deals with perceptual, motor, and cognitive problems, while psychiatry mostly deals with emotional and social problems. The fact that these disciplines are separate is an accident of recent history. If things had gone a little differently in Vienna, St. Petersburg, and Baltimore around the turn of the twentieth century, there might have been a single medical specialty devoted to all brain disease that would integrate both biological and talking-cure therapies. As it is now, there is no biological basis for assigning particular brain diseases to these specialties. It's not as if there is a dividing line in the structure of the brain such that problems of the occipital and parietal lobes are sent to neurologists and those of the temporal and frontal lobes go to psychiatrists. Nor is there a biochemical dividing line: diseases of glutamate-using synapses are not the territory of neurologists, while diseases of dopamine-using synapses go to psychiatrists. In truth, it's just that, like brains themselves, these fields have "evolved," not according to any master plan, but merely in response to the vagaries and constraints of history. Clearly, the perception/emotion distinction cuts deep into the way we think about the brain and the way we deal with its dysfunctions.

What I hope to show here is that perception and emotion are often inextricably linked. There is little, if any, "pure perception" in the brain. By the time we are aware of sensations, emotions are already engaged. Fascinating examples of this can be seen in two complementary types of brain damage. In 1923, the French physician Jean Marie Joseph Capgras described a patient who, following temporal lobe damage, could still visually identify objects and human faces, but these objects and faces did not evoke any emotional feelings. As a consequence, this patient, suffering from what is now called Capgras syndrome, became convinced that his parents had been replaced by exact human replicas. One explanation is that he was led to this conclusion because the emotional responses he expected to feel when seeing his parents weren't there and, consequently, the only reasonable explanation was that these people looked like his parents but were not actually they. The problem was exclusive to vision: the voices of his parents still sounded genuine.

Since the original description, quite a few more cases of Capgras syndrome have come to light and some of these have been observed quite carefully. Capgras syndrome is most often manifest as a feeling of parental imposters, but it can occur for anyone or anything for which there is an expected strong emotional response—pets, for example. Many Capgras patients find mirrors extremely disturbing: they recognize that the reflected image resembles themselves, but they are also convinced that the reflection is of an imposter. Often, this is terrifying because the reflected image is thought of as a malevolent stalker, determined to ruin the life of the patient.

Capgras patients do not have a simple problem with either visual discrimination or emotional responses. In the laboratory, they can easily make distinctions between similar faces and objects. They do not hallucinate and can have appropriate emotional responses to auditory stimuli. These observations, together with anatomical evidence, support the view that Capgras syndrome is specifically a defect in information transfer between the later parts of the visual "what" pathway and the emotion centers, including the amygdala.

The second part of this story about vision and emotion comes from patients who have been blinded by damage to the primary visual cortex. In Chapter 1, I discussed how some patients with this type of lesion can still accurately locate an object in their visual field, even though they have no conscious awareness of seeing anything (blindsight). Recently, a patient who sustained this form of damage from repeated strokes that affected the primary visual cortex was asked to guess the emotions expressed in photos of human faces. These faces, which were both male and female, showed typical expressions of fear, sadness, happiness, and anger. He was able to guess the correct emotion about 60 percent of the time. This was not a perfect score but was significantly better than the outcome obtained by chance. When this task was repeated with the subject in an fMRI machine to scan brain activity, significant activation was seen in the right amygdala for emotional faces, with the strongest activation produced by fearful expressions.

Taken together, these clinical examples show that for both the ancient midbrain visual system and the modern cortical visual system, the amygdala is activated to engage emotional responses. It is likely that, in the case of the cortical "what" pathway, the amygdala is not the only region engaged in the triggering of emotional responses by visual information. The important point here is that visual information is rapidly fed into emotional centers in the brain, which makes it impossible to separate emotion from perception in experience. When an object is rapidly and symmetrically expanding in your visual field, indicating a collision, you cannot help taking evasive action. It's a hard-wired subconscious response. Likewise, when you see a snake in the grass or an angry face, your brain will begin to prepare for "fight or flight" by triggering increased heart rate and other anticipatory physiological responses. This occurs before you are able to consciously make a plan of action. Though the examples I have used are from vision, this principle applies broadly to all of our senses: emotion is integral to sensation and the two are not easily separated.

ONE SENSATION THAT we think of as intrinsically emotion-laden is pain. Pain is not caused merely by the overactivation of sensory pathways in the body, but rather by a dedicated system of sensory cells and their axons that project into the spinal cord, and ultimately to the brain. Counterintuitively, some of the axons that send pain information tend to be of small diameter, and hence they are among the slowest in the nervous system for conducting spikes, operating at a speed of about 1–2 meters per second. This is why, when you stub your toe, you can feel some sensation rapidly (through the fast, nonpain fibers) but you can count a complete "One-Mississippi-Two" before you feel the wave of pain.

Pain is crucial in two ways: it helps protect us from the tissue-damaging effects of dangerous stimuli, and it acts as a warning to learn to avoid these sorts of situations in the future. People who have lost pain sensation owing to nerve damage from trauma or an inherited disease are at constant risk of injury. Pain is not, however, a unitary sensation. It can be divided into multiple components: we now have good evidence that there are separate sensory/discriminative and affective/motivational pathways for pain. The axons of the sensory pathway form synapses in the lateral portion of the thalamus (far from the midline), which in turn send axons to the body representation in the primary somatosensory cortex. Selective damage to this pathway results in a condition in which the ability to discriminate the qualities of pain (sharp versus dull, cold versus hot) is lost. Individuals with this type of lesion may be able to describe an unpleasant emotional reaction to a particular stimulus but are unable to describe its qualities or even specify its location on their body.

The affective (emotional) dimension of pain appears to be mediated by a pathway that runs more or less parallel to the lateral sensory pain pathway: it involves activation of a medial portion of the thalamus (near the midline) that then sends axons to two cortical regions implicated in emotional responses, called the insula and the anterior cingulate cortex. Damage restricted to the medial affective pain pathway results in a condition called pain asymbolia. In this condition people are able to accurately report the quality, location, and relative strength of a painful stimulus, and have intact withdrawal and grimacing reflexes and normal-looking biopsies of their peripheral nerves. What is amazing about people with pain asymbolia, however, is that they seem to lack the negative emotional response to pain that the rest of us take for granted: they can report pain accurately, but it just doesn't seem to bother them. This syndrome can result from a genetic defect (a family in France has been described in which pain asymbolia is inherited) or from traumatic damage to the insula or anterior cingulate cortex.

The affective component of pain can be modulated by cognitive and emotional factors. Anxiety and specific attention to a painful stimulus can increase the affective component of pain while relaxation techniques and distraction can reduce it. A potent behavioral form of pain modulation is hypnotic suggestion that (depending upon the suggestion) can either increase or reduce the affective component of pain. When Catherine Bushnell and her colleagues at McGill University used hypnotic suggestion to increase or reduce the emotional component of pain felt by subjects in a brain scanner, they found corresponding increases and decreases in the activity of the anterior cingulate cortex, which further implicates this structure in a distinct affective pain pathway.

The anterior cingulate may do more than play role in pain; it may also have a more general role in producing emotional responses to tactile stimuli. We have evidence that this structure is also activated by pleasant light touching (caressing, if you will) and may contribute to emotional bonding and hormonal responses evoked by skin-to-skin contact between individuals (I'm not just talking about sex here—the most important example of this may be in parent-child bonding). It may be that different subregions or biochemical pathways in the anterior cingulate are engaged to produce positive or negative emotional responses to tactile stimulation.

Recent experiments on rats have indicated that learning to avoid painful stimuli depends upon the affective/motivational rather than the sensory/discriminative pain pathway. In these experiments, using a simple learning task called conditioned place aversion, a rat is placed in a box with two chambers that are easy to tell apart (typically, one chamber is painted black and the other white). When the rat enters one chamber it receives a moderately painful foot-shock through metal bars in the cage floor. Very quickly, rats will learn to avoid the chamber where the shock was received. But if drugs that block receptors for the neurotransmitter glutamate are injected into the anterior cingulate cortex before training, then conditioned place aversion learning will be blocked. Conversely, if the animals are placed into one chamber and, instead of receiving a footshock, glutamate is injected into the anterior cingulate cortex, they will behave as if footshock had been received and will learn to avoid that chamber. But if these injections, of either glutamate or the receptor blocker, are made into locations in the sensory/discriminative pathway, learning proceeds normally. Thus it is the emotional rather than the sensory response evoked by pain that appears to provide the teaching signal for aversive learning.

Both humans and our hominid and prehominid ancestors live(d) in social groups, so it not surprising that our sensory systems appear to have some particular specializations for social interaction. A recent study has demonstrated that showing people still photographs of hands and feet in painful situations causes activation of brain regions that comprise the affective pain pathway. The anterior cingulate cortex was activated by this experience and its activity was strongly correlated with the participants' ranking of the others' pain. This remarkable finding, that affective pain centers can be activated by both your own painful experiences and those of others, may shed light on the neural substrate of empathy. It will be interesting to repeat these experiments in populations with disorders that impair empathy.

WE OFTEN SPEAK of certain social interactions as being painful or hurtful. Is this only a linguistic metaphor, or do physical and social pain really have a common substrate in the brain? A clever study by Naomi Eisenberger and her coworkers at UCLA has shown that subjects who were made to feel social exclusion in a three-way ball-throwing game showed strong activation of, you guessed it, the anterior cingulate cortex. In order to do this experiment, the ball-throwing game was actually a virtual one using a computer screen visible to the subject in a brain scanner. The exclusion in a virtual ball-throwing game is not a very potent form of social pain, and yet it produced a strong activation of a key affective pain center in the brain. One can only imagine what it would look like if someone in an fMRI machine received a "let's just be friends," phone call from a serious love interest. Considering this study together with the previous one on physical pain empathy, it is reasonable for us to speculate that the anterior cingulate and related structures in the affective pain pathway may be important in empathy for social as well as physical pain.

I have talked at length about how sensory systems in the brain are intertwined with emotion. Now I would like to present the idea that sensation and motor function in the brain are also comingled. Even today, students are shown brain diagrams and told that certain regions of the brain are sensory while others are motor. In truth, in many cases this distinction is not so clear. There are many places in the brain where sensory and motor function are blended, including the cerebellum and the basal ganglia, but now I will focus on one particular example in the cortex with relevance for human social behavior. Several years ago, Giacomo Rizzolatti and his colleagues at the University of Parma made recordings from single neurons in a cortical region of the monkey brain called the ventral premotor area. This region had previously been shown to be involved in the planning of movements. So it was not entirely surprising that certain neurons in this region fired spikes when the monkey performed certain motions, in particular, goal-directed motions such as pushing a button or picking up a peanut to eat. The big surprise came when it was found that a particular neuron that might be activated by the monkey's grasping a cup, for example, would also be activated by watching another monkey performing this same action. These neurons with a dual sensory and motor function were called mirror neurons. It soon became clear that they also could be activated by watching movements of the human experimenter (but not by watching a video of another monkey or human, even when a stereoscopic video was used with 3-D glasses for the monkey). Further investigation showed that mirror neurons can be activated by a wide range of purposeful movements and may be found in other parts of the frontal cortex outside the ventral premotor area.

The discovery of mirror neurons has a lot of brain researchers very excited because this seemingly simple finding holds promise for explaining what have been some very enigmatic issues in human behavior. Humans and, to a lesser degree, the great apes have developed a capacity for understanding the experiences and motivations of others (such as "I know that he knows") that is not present in lower animals. This understanding can be used for good (empathy, cooperation) or evil (manipulation, combativeness) social purposes and has been called a "theory of mind." Mirror neurons, by allowing us to understand the actions of others in terms of our own actions, might be a biological basis of theory of mind. It has even been suggested that mirror neurons may sow the seeds of language in that having a theory of mind is a prerequisite for purposeful linguistic communication: to want to speak you have to have the idea that someone else is listening. We assume, rightly I believe, that mirror neurons are also present in humans. But at the time of this writing that has not been confirmed, because recording from single neurons is humans is a rare procedure and is only done briefly, in conjunction with certain forms of brain surgery.

IN SUM, OUR SENSORY world is anything but pure and truthful. Built and transformed by evolutionary history into a very peculiar edifice, it responds to only one particular slice of possible sensory space. Our brains then process this sensory stream to extract certain kinds of information, ignore other kinds of information, and bind the whole thing together into an ongoing story that is understandable and useful. Furthermore, by the time we are aware of sensations, they have evoked emotional responses that are largely beyond our control and that have been used to plan actions and understand the actions of others.

Chapter Five

Learning, Memory, and Human Individuality

WHAT'S A BRAIN good for? We've seen that the lower portions of our brains have essential control circuits that govern basic body functions: key reflexes, an automatic thermostat, a regulated appetite for food and drink, and wakefulness/sleepiness. The lower brain also has regions for coordinating our movements and modifying our perception to direct our attention to the outside world. This is the basic stuff that we share with frogs and fish, the "bottom scoop of the ice cream cone." The top two scoops, the limbic system and the neocortex, are where things get really interesting. Many complex functions such as language and social reasoning emerge in the cortex, but I contend that there are two key brain functions that are the basis upon which these higher capacities are built. These are memory and emotion—and the interaction between the two.

Consider this analogy: the brain does for the individual what the genome does for the species. The genome, the sequence of information encoded in the DNA, undergoes random mutation and sometimes a mutation (or a collection of mutations) confers an advantage on an individual that allows him or her to have more and/or healthier offspring. The genome, through the Darwinian process of natural selection, is the book in which the story of evolution is written: the experience of the species ultimately modifies the genome and thereby the genetic traits of the species, sometimes rendering it better adapted to the environment. The limitation of evolution through natural selection is that it is not a rapid process. Species adapt to their experiences (environments) slowly, over many generations.

The brain, by storing memories, performs a related function for the individual. It is the book in which individual experience is written. Because memory storage is rapid, it allows an individual to adapt to new experiences and situations. This is a much more flexible and powerful solution than relying solely upon mutation and selection acting upon the genome.

But how does emotion come into it? In our lives, we have a lot of experiences and many of these we will remember until we die. We have many mechanisms for determining which experiences are stored (where were you on 9/11?) and which are discarded (what did you have for dinner exactly 1 month ago?). Some memories will fade with time and some will be distorted by generalization (can you distinctly remember your seventeenth haircut?) We need a signal to say, "This is an important memory. Write this down and underline it." That signal is emotion. When you have feelings of fear or joy or love or anger or sadness, these mark your experiences as being particularly meaningful. These are the memories you most need to store and keep safe. These are the ones that are most likely to be relevant in future situations. These are the building blocks that form logic, reasoning, social cognition, and decision making. These are the memories that confer your individuality. And that function, memory indexed by emotion, more than anything else, is what a brain is good for.

IN CASUAL CONVERSATION we may say that a certain person has a good memory or another person has a bad memory. In truth, however, we know from our everyday experience that things are not so simple. Memory is not a unitary phenomenon. You may have a great ability for matching names to faces while you struggle to memorize music for a piano recital. Your brother might remember everything he ever reads but progresses slowly with motor memory tasks such as learning to improve his golf swing.

Brain researchers have worked for many years to develop a taxonomy of memory, a means of classifying types of memory that has its roots in clinical observation (see Figure 5.1). Much of this work relies on the analysis of human amnesiacs who have sustained damage to various parts of their brains through infections, stroke, trauma, chronic abuse of drugs or alcohol, or, as in the case of the patient called H.M. (Chapter 1), surgery to treat otherwise incurable seizures. Other insights have come from studying more temporary forms of disruption, such as transiently acting drugs and electroconvulsive shock (used to relieve depression that fails to respond to other therapies).

In the 1950s it was generally thought that patients like H.M. and others who sustained damage to the hippocampus and surrounding cortical tissue were unable to form any new memories at all. But detailed study of these patients revealed that although they could no longer form new memories of facts or events, so-called declarative memories, they could lay down memory traces for a number of other tasks. One of these is mirror reading: learning to read words in English that have been printed with left-right reversal (Figure 5.2). This is a task that both normals and hippocampal amnesiacs such as H.M. can learn with daily practice. It's also a nice task for illustrating different types of mem-

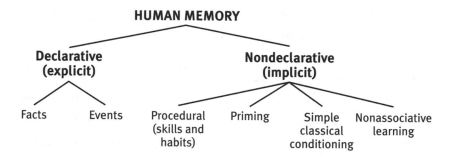

FIGURE 5.1. A taxonomy of human memory. Adapted with permission from Elsevier from B. Milner, L. R. Squire, and E. R. Kandel, Cognitive neuroscience and the study of memory, *Neuron* 20:445–468 (1998). *Joan M. K. Tycko, illustrator.*

ory: although both the amnesiacs and the normals showed daily improvement in the mirror-reading task (as indicated by progressively faster reading times), only the normals could recall some of the words used in the test the previous day—the amnesiacs had no memory of these words whatsoever (indeed, they also had no memory that the previous day's training session had even occurred).

Further experiments with hippocampal amnesiacs have revealed a large group of memory tasks that are retained. Amnesiacs still have memory for motor co-ordination—they can improve at sports with practice. Both mirror reading and motor coordination learning fall into the larger category of "skills and habits" shown in Figure 5.1. Amnesiacs also retain the ability to learn simple, subconscious associations, through a process called classical conditioning. For example, your heart rate will reflexively accelerate if you receive a mild shock to your arm, but it will not do so in response to a more neutral stimulus such as the sight of a dim red light briefly appearing in your field of view. But if the light is paired with the shock repeatedly, after a while your brain will begin to learn that the light predicts the shock and your heart rate will accelerate in response to the

Kludge
Jacob
cerebellum
Natalie

FIGURE 5.2. Mirror reading, a skill that both hippocampal amnesiacs and normals can acquire and retain with practice. The memories of the particular words read, however, will be retained only by the normals.

light alone. Hippocampal amnesiacs trained in this task for several days will have no memory of the previous day's training, but their heart rate will accelerate in response to the light nonetheless.

Perhaps the most interesting form of memory retained in amnesiacs is achieved by a method called priming. In this task, initially devised by Elizabeth Warrington and Larry Weiskrantz of Oxford University, amnesiacs are asked to recall a list of words they have seen the previous day. Not surprisingly, if you simply ask them to list the words from the earlier session they have no memory of them at all. But if you give them the first few letters of a word, they will often be able to correctly produce the complete word even if it feels to them as if they are guessing randomly. For example, if a word on the list was "crust," the stem "cru____" would probably evoke the correct answer rather than other possibilities such as "crumb" or "crud" or "cruller." What's interesting about priming is that, unlike many of the other forms of memory retained in hippocampal amnesia, it is a cognitive rather than a motor task.

All of these memory tasks that are retained in amnesiacs (priming, skill and habit learning, classical conditioning, as well as some others I didn't discuss) fall into a category called nondeclarative, or implicit, memory: they are forms of memory that do not involve conscious retrieval. These memories are not recalled, but rather are manifest as a specific change in behavior. Nondeclarative memory is *not* what we usually think of when we talk about memory in casual conversation—it is not memory for facts and events—such as what you had for breakfast yesterday morning or the name of the British prime minister. Nonetheless, nondeclarative memory is central to our experience.

OBSERVATIONS OF HUMAN amnesiacs clearly suggest that storage of new declarative memories requires an intact hippocampal system (the hippocampus proper and some adjacent cortical structures). This brings up a central question: are particular memories stored in specific locations in the brain or are they stored in a distributed fashion, spread over many brain regions? One early indication of the answer came from the work of the Montreal Neurological Institute neurosurgeon Wilder Penfield, who, starting in the 1930s, stimulated the brains of patients undergoing surgery for epilepsy.

This was not just an academic exercise. It allowed him to map more carefully than previously the location of the area that triggered the seizure and thereby minimize damage to nearby parts of the brain. Because brain tissue has no pain-sensing system itself, neurosurgery can be performed on conscious people while they are under a local anesthetic to block pain from their scalp and

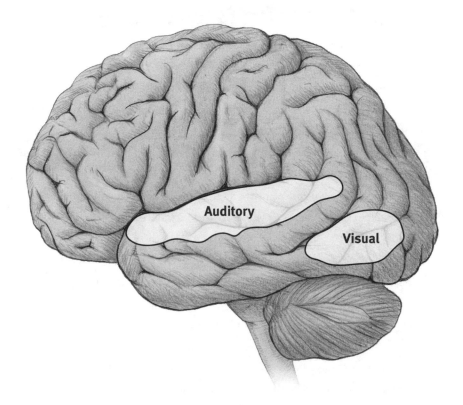

FIGURE 5.3. Reminiscence evoked by brain stimulation during neurosurgery. The Canadian neurosurgeon Wilder Penfield inserted electrodes to stimulate the cortical surface of awake patients in the course of neurosurgery. This figure shows two types of memory-like experiences evoked by stimulation in various regions. *Joan M. K. Tycko, illustrator.*

skull. Penfield's stimulation was restricted to the cortical surface and was performed on over 1,000 surgical patients (Figure 5.3). In a small fraction of cases, stimulation of the cortical surface would evoke a coherent perception: a snatch of music, a human voice, a vision of a pet or loved one. Were these electrical stimuli evoking memories? Well, yes and no. In some cases it does seem that particular, real past events were recalled, at least in fragmentary form. More often, however, the stimulation evoked sensations that were dreamlike, with typical elements of fantasy and violations of physical laws. Often, the area that evoked a "memory" was itself the epileptic focus. In these cases, destruction of that cortical tissue did not obliterate the memory of that particular stored experience. So, the Penfield experiments, while titillating, did not directly address the question of memory localization.

If formation of new nondeclarative memories can proceed when the hippocampal system is destroyed, then where are the critical locations for these forms of memory? Some information about this can be derived from studies of humans with damage to other brain regions. Damage to the amygdala, for example, seems to be associated with memory storage for classical conditioning of emotional responses, particularly fear conditioning. Damage to the cerebellum has similar effects for classical conditioning of emotionally neutral stimuli (see Figure 5.4).

So, is memory storage localized or not? The answer is not so simple. It's also a bit different for nondeclarative versus declarative memory. Nondeclarative memory is not consciously recollected. Rather, it is evoked by a specific stimulus or set of stimuli and is manifested as a change in behavior. As a result, nondeclarative memories can often be localized, not just to a brain region, but to a certain subregion or even class of neuron. Declarative memory is a different story. Such memories are consciously recollected. They are useful in large part because we can access them using stimuli that are very different from the ones that created them initially. For example, when you read "Imagine your mother's face," the sensation of reading that line is nothing like the stimuli that laid down the memory of your mother's face, yet you can probably re-

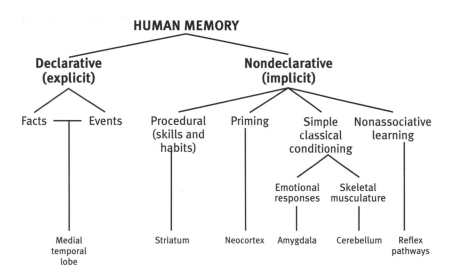

HUMAN MEMORY

Declarative (explicit)
- Facts — Events
  - Medial temporal lobe

Nondeclarative (implicit)
- Procedural (skills and habits)
  - Striatum
- Priming
  - Neocortex
- Simple classical conditioning
  - Emotional responses
    - Amygdala
  - Skeletal musculature
    - Cerebellum
- Nonassociative learning
  - Reflex pathways

FIGURE 5.4. A taxonomy of human memory, now elaborated to show some crucial brain regions involved in different tasks. Here, the medial temporal lobe means the hippocampus and some associated regions of the cerebral cortex. It should be cautioned that, in real-world behavior, most experiences will simultaneously lay down memory traces of several types. For example, if you take lessons to improve your tennis game, you will probably recall particular things that may have happened during the lesson (declarative memory for events), but what you are really trying to achieve is an unconscious improvement of your motor performance as you play (nondeclarative skill memory). Adapted with the permission of Elsevier from B. Milner, L. R. Squire, and E. R. Kandel, Cognitive neuroscience and the study of memory, *Neuron* 20:445–468 (1998). *Joan M. K. Tycko, illustrator.*

call her face with ease after reading that line. This imposes an important constraint on declarative memories: nondeclarative memories can simply be accessed in a subconscious fashion through specific stimuli, but declarative memories must be embedded in a much richer informational system, which makes it less likely that they will be localized to the same degree as nondeclarative memories.

ALTHOUGH DAMAGE TO the hippocampal system produces anterograde amnesia, an impairment in the storage of new memories for facts and events, it does not erase a lifetime's worth of declarative memories. Rather, there is typically a "hole in declarative memory," or retrograde amnesia, which stretches back 1 or 2 years before the infliction of hippocampal damage. Thus H.M. and others like him have lost a part of their past forever, but older memories have been spared. The explanation for this seems to be that declarative memories are initially stored in the hippocampus and some adjacent regions, but, gradually, over months to years, the storage site changes to other locations in the cerebral cortex. The dominant theory today is that the final locations for declarative memories are distributed in the cortex, not in a random fashion, but rather in those parts of the cortex initially involved in their perception. In this fashion, memories for sounds are stored in the auditory cortex (indeed, memories for words appear to be stored in a particular subregion of the auditory cortex), memories for scenes in the visual cortex, and so on. What this means for any real experience, involving multiple senses, is that your memory for, say, your first trip to the beach is stored in a number of cortical locations, each corresponding to a particular sensory modality or submodality. There does not appear to be a single dedicated site for the permanent storage of declarative memories. This underlies, at least in part, the observation that memory is not a unitary system. This may be why your Aunt Matilda can remember every word to every song Elvis Presley recorded, but can't keep track of your birthday.

MEMORIES MAY BE classified not only by type but also by duration. There is evidence for separate neural processes underlying at least three stages of memory. The first and most transient of these is known as working memory. Anyone who grew up with an annoying sibling knows certain aspects of working memory well: You've just read a phone number from your address book and you're repeating it to yourself, trying to keep it in your memory long enough to dial the phone while your sib is trying equally hard to interfere by shouting random numbers in your ear. Working memory is a temporary "scratchpad" for holding information just long enough to complete a task (to dial a phone number or to remember the first part of a heard sentence long enough to match it with the ending). You can hold information in this scratchpad for a somewhat longer time through rehearsal or by employing mental imagery, but otherwise it will quickly fade away. Working memory is a form of declarative memory that is crucial to understanding lengthy experiences as they unfold over time. It is the glue that holds our perceptual and cognitive lives together.

Working memory is preserved in hippocampal amnesiacs. Although we don't have a complete understanding of its neural basis, there is now a generally accepted model that holds that working memory requires the ongoing firing of particular sets of neurons. This has been tested in monkeys by using a working-memory task called delayed matching to sample (Figure 5.5). In this task, a colored light briefly flashes, and then after a delay of a few seconds the monkey must correctly choose the previous color from a display of two or more to get a food reward. Investigators found that some neurons in the higher regions of the visual "what" pathway (in an area called TE) fired continually during the working-memory interval. These higher visual areas send a lot of axons to the prefrontal cortex, and neurons in this area also fire in this fashion. Similar activity can also be recorded with scalp electrodes in the prefrontal cortex of human subjects performing working-memory tasks. So, a current model is that there are separate working-memory systems for different areas in the brain, each located at some point in the appropriate region of cortex (auditory, visual,

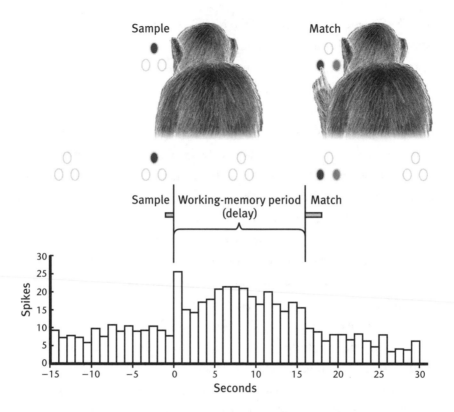

FIGURE 5.5. Persistent neuronal activity as a substrate of working memory. In this delayed matching to sample task, a monkey sees a given color and then, after a delay of a few seconds, must pick this color from two choices. The lower panel shows a recording made from a neuron in the higher visual region called area TE, illustrating that the firing rate of this neuron was elevated when the sample was presented and remained so through the 15-second-long working-memory interval. Adapted from L. R. Squire and E. R. Kandel, *Memory: From Mind to Molecules* (Scientific American Library, New York, 1999); © 1999 by Scientific American Library; used by permission of Henry Holt and Company, LLC. *Joan M. K. Tycko, illustrator.*

and so on). These regions all seem to project to the prefrontal cortex, which, at least to some degree, integrates working memory across sensory modalities. This model is further supported by the findings that damage to the prefrontal cortex in both humans and monkeys results in impairment on working-memory tasks.

More subtly, if the prefrontal cortex of monkeys is artificially electrically stimulated during the working-memory interval, performance can be disrupted. A similar effect can be produced by injecting drugs into the prefrontal cortex that either block or overactivate receptors for the modulatory neurotransmitter dopamine. Dopamine functions to tune the amount of spike firing in the prefrontal cortex that is triggered by information flowing from other cortical regions such as the auditory and visual systems. This may explain why schizophrenics and patients with Parkinson's disease, people whose ailments are associated with defects in dopamine signaling, perform poorly on tests of working memory.

IF YOU QUERY middle-aged people on general knowledge (news, popular culture), you typically find that they have better recollection of recent events than more distant past ones. This predictable result is called the forgetting curve. Yet distant memories that do survive normal forgetting are unusually resistant to disruption. As the record of a particular experience moves from working memory through short-term memory and into long-term memory, the memory trace, or engram (those changes in the brain that encode memory), gradually changes from being fragile and easily disrupted to being more stable. This transformation process takes time and has been given the name consolidation. The evidence for this comes from both human and animal studies. If you repeat the experiments mentioned above in which you query general cultural knowledge among people who have received bilateral electroconvulsive shock treat-

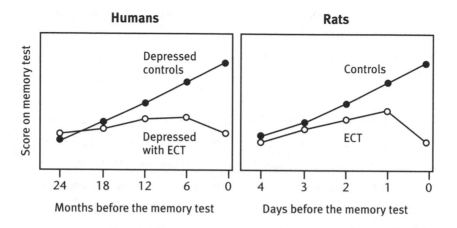

FIGURE 5.6. The persistence of old memories and the fragility of new memories. Both humans and rats were tested for their recollection of past events. Controls showed some degree of forgetting of information in the more distant past. Both humans and rats receiving ECT had severely impaired memory for events occurring immediately before ECT, but had normal rates of forgetting for information in the more distant past. Note the different time scales for the human and rat data. *Joan M. K. Tycko, Illustrator.*

ment (ECT) to relieve drug-resistant depression, you will find that superimposed upon the forgetting curve is an additional disruption of memory that is strongest for events that occurred immediately before the treatments and that gradually trails off as you query events further in the past (Figure 5.6). Of course, in this type of experiment, it is important that the control group be others with severe depression and not just the general population.

Similar studies can also be done using laboratory animals such as rats. Obviously, in this case, you can't quiz them about general knowledge, so instead you train them in a particular task (such as navigating a maze to get food) and then wait for various intervals before giving ECT. When you test them for their memory of the maze task the next day, you will find a result that parallels that seen in humans: recent memory is easily disrupted by ECT, but memory in the more distant past is sufficiently consolidated to withstand this treatment (Figure 5.6). This basic strategy can be applied to many types of experimental amnesia, including those caused by the administration of various drugs. One class of drug that has been particularly well studied in terms of disrupting memory consolidation is protein synthesis inhibitors. These are compounds that interfere with any one of several biochemical steps by which genes ultimately direct the synthesis of new proteins. Thus a popular hypothesis is that synthesis of new proteins is one important step in the consolidation of short-term into long-term memory, thereby rendering the memory less vulnerable to erasure. Although these examples use tests of declarative memory, several lines of evidence indicate that nondeclarative memories also undergo consolidation and that this consolidation requires new protein synthesis.

ON THE EVENING of October 6, 1996, I was watching TV. I know this, because so were about 46 million others in the United States: It was the night of the first Presidential debate between incumbent Bill Clinton and challenger Bob Dole. When questions were posed to Clinton, he had a habit of pausing for about 3 seconds with his eyes rolled back in his head and then launching into a carefully constructed and detailed answer. However you might have felt about his policies, you had to admire his command of information. After several questions like this, with the characteristic 3-second pause, my wife said, "Look, he's rewinding the tape!" We laughed because it really did seem as if something mechanical was happening in the President's brain that night.

Although we might imagine that our memories for facts and events are stored on a tape we can rewind or a set of photographs we can browse in an album, this does not seem to be true. As discussed previously, one of the biggest challenges of declarative memory (memory for facts and events) is to store information so that it can be retrieved by diverse stimuli. The key point here is that retrieval of memory is an active process. It is not like browsing through an album of photographs, even an album of fading photographs. Rather, it is a bit like searching the Internet with Google. A question such as "Who was with us on that day trip to the beach last summer?" provides a few search terms that will yield a large number of memory fragments associated with key terms such as "beach" and "last summer." But the question "Who was with us on that day trip to the beach last summer—you know—when we got caught in the thunderstorm and then you threw up in the car on the way home?"—with its greater number of search terms—not only makes it more likely that you will recall the memory of those events, but also makes it more likely that you will recall more aspects of the events. Of course, unlike a Google search, retrieval of declarative memory is not fundamentally text based.

Memory retrieval is an active and dynamic process. But this dynamic recollection and rewriting of memory is a two-edged sword. In some ways it's very useful for subsequent experience and recollection to modify memory traces of events in the past, but this can also lead to errors. Memories of recurring commonplace events are often rendered generic. This is something we all know from our own personal experience. As a child growing up in Santa Monica, California, I probably ate dinner with my father at Zucky's Delicatessen hundreds of times. Although I have many memory fragments associated with those times—the smell of matzo-ball soup, my father's secret-agent like insistence upon sitting with a view of the door, the weird mechanical sound of the cigarette machine, the unnatural colors of the glossy marzipan fruits in the bakery case—these are mostly not related to specific incidents. I can't really say that I remember a particular meal that I ate any night when I was 12 years old. I can, however, remember almost everything about the particular night at Zucky's in 1974 when my father told me that he would be having triple bypass heart surgery. It scared the hell out of me and, as a consequence, that particular meal is etched into my memory forever.

Everyone knows how emotion-laden events can be written into long-term memory with unusual strength. One would be tempted to think that this could be entirely explained by activation of emotional systems at the time of the event. Indeed, that's part of the story, but it's not the whole story. It is now clear that consolidation of long-term memory is also reinforced by subsequent conversation—when you repeatedly tell the story of where you were on 9/11, this repetitive narration reinforces consolidation. Furthermore, the emotions in both you and your listeners that are evoked by the retelling will subtly influence the memory trace itself—the event and the retelling will begin to blend in your mind.

This dynamic reconsolidation of memory is all well and good in some ways: memory of commonplace events is probably of more use to us when rendered generic by the passage of time and subsequent experience. This also has the useful result that emotionally important events stand out more clearly in our memories. But this dynamic process renders our memories particularly subject to certain forms of error above and beyond the slow, gradual fading of long-term memory over time. In his splendid book *The Seven Sins of Memory*, the Harvard University psychologist Daniel Schacter speaks of three of these "sins of commission" in declarative memory retrieval: misattribution, suggestibility, and bias.

Misattribution is very common form of error in which some aspects of a memory are correct but others are not. It can happen in many domains. Take, for example, source misattribution: I may correctly recall a joke I heard which begins "Ted Kennedy walks into a bar . . ." but I will swear I heard it from my sister-in-law when I really heard Jay Leno tell it on TV. Sometimes misattribution can cause you to think that you've created something original, whereas really you had heard it from another source and attributed it to your own internal processes. I went around for over 30 years humming a snippet of tune I thought I had composed myself, only to hear it years later when I bought my children a DVD containing Bugs Bunny cartoons from the 1940s.

Misattribution is at the heart of one of the most famous cases in music copyright law, which involved the 1970 number-one pop hit by George Harrison called "My Sweet Lord." Although the lyrics and instrumentation differ, the tune of "My Sweet Lord" very strongly resembles that of a previous number-one hit recorded by the Chiffons in 1963 called "He's So Fine." The judge in the case ruled that though Harrison had no intent to plagiarize, he had almost certainly misattributed his memory of the tune of "He's So Fine" (which Harrison admitted he had heard before), thereby imagining that he had composed it de novo. The company which held the copyright to "He's So Fine" was ultimately awarded millions of dollars in damages from Harrison.

These examples are forms of source misattribution. A variant of this is misattribution of time or place. A common experimental design is to give subjects a list of words to study. When they return the next day they are given a new list of words and are asked to indicate which ones they had seen the day before. People in these experiments will often misattribute new words to the previous list. Their propensity for doing this can be manipulated by experimental context. For example, if a word appearing on the new list for the first time is more familiar to the subject or is thematically related to several words on the first list, it is more likely to be misattributed. If the first list contained "needle," "sewing," "pins," and "stitch," then the chance of misattributing the word "thread" to the first list will be high. It may be that we have an evaluative system in our brains that says "If I recognize this word rapidly then it's likely that I've seen it before," and this is the basis of some forms of misattribution.

Suggestibility and bias are additional forms of memory error in which the act of recollection involves the incorporation of misleading information. Suggestibility is the term used when this information comes from external sources (other people, films, books, media), while bias is warping one's recollection to fit present circumstances: "I always knew that the Red Sox could win the World Series." It turns out to be surprisingly easy to alter people's recollections. For example, a number of studies have sought to simulate police line-ups: a group of experimental subjects will watch a video of a (simulated) convenience store robbery and will then see a line-up of six suspects, none of whom was the robber in the video. When subjects are presented with the suspects one by one and asked to make a yes-or-no decision, almost all will correctly respond "no" to all six. But if the six are presented all at once and the subjects are asked, "Is any of these the robber?" then about 40 percent of people will pick a suspect (usually the one who resembles the perpetrator most closely). If the subject is told by the experimenters in advance that several others have already identified suspect X and they need them to confirm or deny, then about 70 percent of people can be manipulated into false recollection. These results not only highlight the suggestibility of memory recall but also have obvious implications for police procedures and our legal system.

The problem of suggestibility is even greater in children, particularly preschool-age children. In a typical study, a group of preschoolers had a bald man visit their room, read a story, play briefly, and leave. The next day, these children were asked nonleading questions such as "What happened when the visitor came?" and they related a series of memories that, though not complete, were quite accurate. But when leading questions were used, such as "What color hair did the man have?" then a large number of children made up a color. Even those children who initially responded that the man had no hair would typically, after having the question repeated several times in different sessions, begin to confabulate and even extend the false recollection—"He had red hair.

And a mustache too!" Initially, such studies were done using rather innocuous questions like the one above. The consensus at that time was that though children were suggestible about trivial details, they could not easily be made to confabulate entire events, particularly events that would be emotionally traumatic.

A series of high-profile accusations of child abuse in the 1980s prompted several teams of researchers to reexamine this point. What they found was startling. Both preschoolers and, to a lesser extent, elementary school children could easily be made to completely manufacture allegations of abusive behavior (such as yelling, hitting, or taking off their clothes) against an adult in a laboratory setting. All it took was some social incentives: leading questions, reinforcement of particular answers, and a lot of repetition. These are exactly the techniques that were used by many therapists and police officers in developing evidence to accuse preschool teachers in the 1980s. Most (but not all) of these cases were ultimately dropped or overturned on appeal. Let's be clear about what this means: Abuse of children happens and spontaneous reports of abuse volunteered by children are often true and warrant careful examination. But extreme care must be taken in questioning children in cases where abuse is suspected. It is extraordinarily easy for caring professionals with the best intentions to distort a child's recollection or even implant memories that are completely false. The neural basis for the increased suggestibility of small children is unknown but is likely to reflect the fact that brain regions required for retaining memory of events and evaluating confidence in the accuracy of one's own recollections, particularly the frontal lobes, are still undergoing rapid growth and reorganization in the preschool years and slower growth from age 5 to age 20.

WHAT CHANGES OCCUR in brain tissue to store long-term memory? Let's begin our consideration of this key question by stepping back a bit and playing engineer. In building neural memory storage there are a lot of difficult design goals we'll have to meet. First, the capacity for memory storage must be large. Even though we forget things, we still have to store a huge amount of information over many years and do so with reasonable fidelity. Second, memory must be durable. Some memories will last for an entire lifetime. Third, memories must be stored in a way such that they are retrieved readily, but not too readily. For declarative memories, this means that they must be recollected by using fragmentary cues that can be very different from those which laid them down ("Imagine your mother's face"). Nondeclarative memories are optimally triggered by an appropriate range of stimuli—if you've been trained to blink to a 400 hertz tone, then you probably also would want to blink to a 410 hertz tone but not a 10,000 hertz tone. Fourth, memories must be malleable, based upon subsequent experience in order to place them in a useful context and absorb them into the totality of the conscious self. All in all, this is a rather tall order. Memory must be accurate, but it must also be useful in supporting generalization. It must be permanent, but also subject to modification by subsequent experience. Given these competing requirements, it is not surprising that our memories for facts and events are often subject to misattribution, suggestibility, and bias.

On a smaller scale, what we need to build are systems through which particular patterns of experience-driven neuronal activity will create enduring changes in the brain. What are the general classes of change that could be used to store memories? We know that the fundamental unit of neuronal information is the spike. The probability of spike firing is driven by the integrated activity of many of the excitatory and inhibitory synapses, which add together to produce changes in the voltage across the membrane at the axon hillock, where the spike originates. So if a particular pattern of neuronal activity results in a lasting modification of, say, voltage-sensitive sodium channels located at the axon hillock, such that the threshold for firing a spike was moved closer to the resting

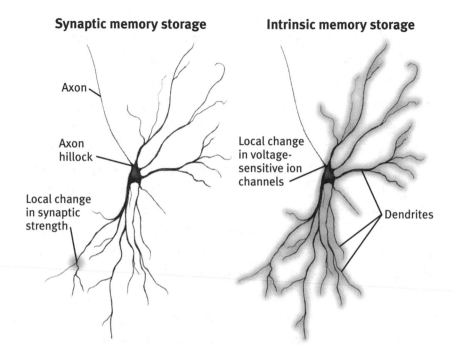

Synaptic memory storage

Axon

Axon
hillock

Local change
in synaptic
strength

Intrinsic memory storage

Local change
in voltage-
sensitive ion
channels

Dendrites

FIGURE 5.7. Synaptic versus intrinsic modulation in memory storage. Long-term
modulation of synaptic strength (left) results in changes in throughput
that are confined to the activated synapses (shaded area). Changes in intrinsic excitability through modification of voltage-sensitive channels in
the axon hillock (right) will change throughput from synapses received
throughout the dendritic arbor (shaded area). As a consequence, intrinsic
changes have the advantage of producing useful generalization but the
disadvantage of having a much smaller capacity to store memory. Adapted
from W. Zhang and D. J. Linden, The other side of the engram: experience-driven changes in neuronal intrinsic excitability, *Nature Reviews
Neuroscience* 4:885–900 (2003). *Joan M. K. Tycko, illustrator.*

potential, then this could produce a lasting change in the firing properties of that neuron, thereby contributing to an engram. This is only one of many possible changes that would affect neuronal spiking. For example, modifying the voltage-sensitive potassium channels that underlie the downstroke of the spike could change their average time to open. This would result in alterations to the rate and number of spikes fired in response to synaptic drive. Indeed, changes in voltage-sensitive ion channels can persistently alter the intrinsic excitability of neurons and, in animal experiments, these changes can be triggered by learning.

Although changes in intrinsic excitability are likely to contribute to some aspects of memory storage, it's unlikely that they are the whole story. Computationally, this mode of memory storage doesn't make the most efficient use of the brain's resources. Recall that there are about 5,000 synapses received by the average neuron. When you change ion channels underlying spike firing, you are changing the probability of firing a spike in response to synaptic input for all 5,000 of those synapses at the same time. One can imagine that this generalizing property might be useful for certain aspects of memory storage, but an engram solely built upon modifying neuronal intrinsic excitability would, by its nature, have a much smaller capacity than one that allowed individual synapses to change.

Experience-dependent modification of synaptic function is a general mechanism that is thought by most brain researchers to underlie a large part of memory storage. There are many steps in synaptic transmission and several of these are subject to long-term modulation. As a sort of shorthand, people speak of "synaptic strength" as a parameter that can be changed. If, as a test, you stimulate 10 excitatory axons to fire spikes and they all converge on the same postsynaptic neuron and you then measure the resultant deflection in membrane voltage (the EPSP), you might find that this produces a depolarization of 5 millivolts. If, after a certain period of conditioning stimulation (a particular pattern of activation designed to mimic the results of sensory experience), this same test stimulation produced a depolarization of only 3 millivolts, this would be called synaptic depression. An increase in the response to 10 millivolts would be called synaptic potentiation. If these changes were long-lasting in nature, they could contribute to the storage of memory. Because there are about 500 trillion synapses in your brain, this mechanism, experience-driven persistent changes in synaptic strength, has a very high capacity for information storage.

There are two general ways to modify the strength of existing synapses. On the presynaptic side, you could potentiate or depress the amount (or probability) of neurotransmitter release following arrival of an action potential. Or, on the postsynaptic side, you could potentiate or depress the electrical effect produced by a constant amount of released neurotransmitter. In molecular terms, each of these forms of modification can come about in several different ways. For example, if one modifies voltage-sensitive calcium channels in the presynaptic terminal so that they pass fewer calcium ions into the cell when an action potential invades, this will depress neurotransmitter release. A similar effect may be produced by modifying the proteins that control the fusion of neurotransmitter-laden synaptic vesicles with the presynaptic membrane. In this case, for a constant spike-evoked presynaptic calcium signal the probability of a vesicle being released would become lower. On the postsynaptic side, you can depress the effect of released transmitter by reducing the number of neurotransmitter receptors in the postsynaptic membrane. Alternatively, a similar result could be achieved by keeping the number of receptors constant but modifying them so that they pass fewer positively charged ions when they open. The point here is that almost every function on both sides of the synapse is subject to modulation and is therefore a candidate for a memory mechanism. In prac-

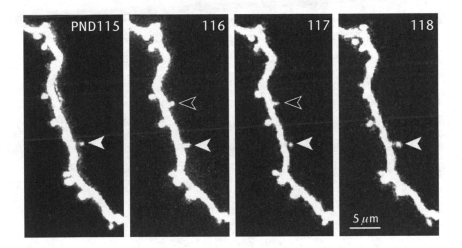

FIGURE 5.8. Changes and stability in the fine structure of dendrites in adult cerebral cortex. These images of a segment of neuronal dendrite from living mouse visual cortex were taken every day from day 115 to day 118 (PND = postnatal day). The filled arrowhead shows one of several stable dendritic spines, while the open arrowhead shows a transient one. This mouse was genetically engineered to express a fluorescent protein in some of its cortical neurons. Reproduced with the permission of Elsevier from A. J. Holtmaat, J. T. Trachtenberg, L. Wilbrecht, G. M. Shepherd, X. Zhang, G. W. Knott, and K. Svoboda, Transient and persistent dendritic spines in the neocortex in vivo, *Neuron* 45:275–291 (2005).

tice, these different molecular mechanisms are not mutually exclusive, and in most synapses several can be working at the same time.

Modifying synaptic function is not the only way to create long-term memories. Such memories may also be encoded through changes in synaptic structure. Although the overall wiring plan of the brain is largely fixed in adult brains, the same cannot be said of individual axons, dendrites, and synapses. Short-term memory is likely to involve changes in the function and structure of existing synapses, but long-term memory can involve the creation of new branches of dendrites and axons. The tiny spines that cover dendrites are structures that are particularly subject to experience-dependent rearrangement. One recent study by Karel Svoboda and his coworkers at Cold Spring Harbor Laboratory used a novel form of microscopy to repeatedly examine dendritic structure in the cerebral cortex of living adult mice (Figure 5.8). They found that over a period of 30 days, approximately 25 percent of dendritic spines disappeared or were newly formed. At a microscopic level, the synapses of the brain are not static. They grow, shrink, morph, die off, and are newly born, and this structural dynamism is likely to be central to memory storage.

I'VE NOW PRESENTED a theoretical overview of some cellular mechanisms by which memory could be stored in the brain. How do we then go about testing whether any of these mechanisms really operate in behaving animals? There are two general approaches. One is to alter the brain function (with drugs, lesions, genetic manipulation, electrical stimulation, and so on) and observe the resultant effects on behavior. This is an interventional strategy (in animals at least; in humans we usually let nature make the lesions). The other is a correlational strategy, where we measure physiological properties of the brain (electrical activity, microscopic structure, biochemistry, gene expression, and so on) to try to determine how they change as a result of experience.

To get a sense of the current state of the struggle, let's examine a form of declarative memory for which scientists have made substantial progress in illuminating the cellular and molecular substrates of the engram. After reports of the amnesiac patient H.M. became known in the 1950s, there was a determined effort to reproduce his deficit, complete anterograde amnesia for facts and events, in an animal model (preferably an inexpensive animal like a rat). It wasn't until the 1970s that this really started to pay dividends. It's not hard to use surgical techniques to damage the hippocampus of a rat. The challenge was to find appropriate declarative memory tasks for this animal. The best ones turned out to be tests of spatial learning.

There are several ways to test spatial learning but the most widely adopted have had animals learn to navigate a maze in order to escape from a situation they find stressful. One clever maze was developed by Richard Morris and his colleagues at the University of Edinburgh. It's not what we normally think of when we hear the word "maze." This maze has no passageways. Rather it consists of a circular swimming pool 1.2 meters in diameter with a wall at the edge to prevent escape and dry milk powder added to make the water opaque. The pool is housed in a room with prominent and unique visual landmarks on the walls to aid in navigation. A rat (or mouse) is placed at a random location at the edge of the water and is then allowed to explore by swimming. Eventually, it will find that there is an escape platform, the top surface of which is just a centimeter or so below the opaque surface of the water. When the rat reaches the platform, it is allowed to stand there for a moment before being gently returned to its cage. The task is to remember where this escape platform is located so that on subsequent trials the rat can swim to it directly and make a quick exit. Not surprisingly, rats that have had their hippocampus surgically destroyed on both sides of the brain cannot learn the Morris water-maze task. Even after many trials, they behave as if they are experiencing the maze for the first time. This appears to be a specific deficit in spatial memory because they can easily learn to swim rapidly to a platform marked with a flag, which indicates that they do not merely have a problem with swimming or vision but rather have a genuine and specific memory deficit.

Also in the 1970s, a report on hippocampal physiology fired the imagination of brain researchers around the world. Terje Lomo of the University of Oslo and Tim Bliss of the National Institute of Medical Research in the United

Kingdom reported that if they briefly stimulated a population of glutamate-using excitatory synapses in the hippocampus of anesthetized rabbits at high frequency (100 to 400 stimuli per second for 1 or 2 seconds), this produced an increase in synaptic strength that could last for days. This phenomenon was named long-term synaptic potentiation (commonly abbreviated LTP). You can see why people got so excited. LTP was an experience-dependent, long-lasting change in neuronal function that occurred in a location in the brain already known to be crucial for memory. Furthermore, high-frequency bursts of the kind known to trigger LTP occur naturally in rats (and rabbits and monkeys). The hypothesis that LTP might underlie memory storage for facts and events in the hippocampus rapidly became one of the most exciting and controversial ideas in brain research.

In the years that followed, thousands of papers were published about LTP. One of the most interesting things scientists learned is that, although LTP was initially found in the hippocampus, it is actually a phenomenon that occurs throughout the brain. It is found in the spinal cord and in the cerebral cortex and almost everywhere in between. Although it is most commonly studied at excitatory synapses that use glutamate as a neurotransmitter, it is present in other types of synapse as well. Another important finding is that there is a complementary process: a persistent use-dependent weakening of synapses called long-term synaptic depression, or LTD. The exact parameters for evoking LTP and LTD vary from synapse to synapse, but at most locations LTP is produced by brief, high-frequency activation (100 stimuli per second for 1 second is typical) while LTD is produced by more sustained activation at moderate frequencies (say, 2 stimuli per second for 5 minutes). So far, all synapses that have LTP also have a form of LTD and vice versa.

Given that random, low-frequency spiking of neurons goes on all of the time, how does the synapse undergo LTP when there's a burst of high-frequency stimulation, but not in the presence of ongoing background activity? This is a problem the brain has solved using several different molecular strategies. Here, I'll consider the most commonly used solution, which involves a special receptor for the neurotransmitter glutamate.

Previously, I've talked about glutamate receptors that rest with a closed ion channel and then open this channel when glutamate binds, allowing the sodium ions to flow in and potassium ions to flow out. This type of glutamate receptor is called an AMPA-type glutamate receptor (named after a synthetic drug that activates it strongly). These receptors cannot differentiate low-level background activity from high-frequency bursts. They are activated by both stimuli. The receptor that can make this differentiation (also named after a potent synthetic drug) is the NMDA-type glutamate receptor (Figure 5.9). The reason that the NMDA receptor can perform this trick is that, at the resting potential of −70 millivolts, its ion channel is blocked by a magnesium ion from the outside (magnesium ions float freely in the saltwater solution that surrounds neurons). This blockade remains until the membrane potential becomes more positive than about −50 millivolts.

So, neither glutamate binding alone nor depolarization of the membrane alone will open the NMDA-type receptor's ion channel. Background activity will produce the former but not the latter, but bursts of high-frequency spikes will produce both glutamate binding and depolarization and the ion channel will open. This ion channel is also unique in that it allows the influx of calcium ions together with sodium ions, while most AMPA-type receptors allow only sodium influx. This means that strong calcium influx through NMDA-type glutamate receptors is a unique consequence of high-frequency bursts. Or, stated another way, the NMDA receptor is a coincidence detector: it opens and fluxes calcium ions when both glutamate is released *and* the postsynaptic membrane is depolarized, but neither of these events alone is sufficient.

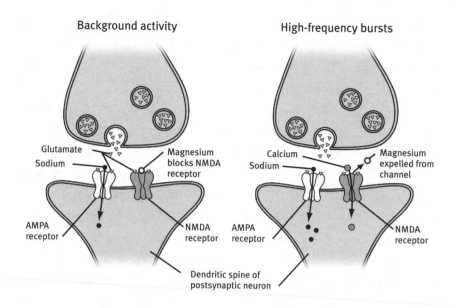

Background activity　　　　　High-frequency bursts

Glutamate　　　　　Magnesium　　　Calcium　　　　　Magnesium
Sodium　　　　　　blocks NMDA　　Sodium　　　　　expelled from
　　　　　　　　　receptor　　　　　　　　　　　　channel

AMPA　　　　　　　　　NMDA　　AMPA　　　　　　　　　NMDA
receptor　　　　　　　receptor　receptor　　　　　　　receptor

Dendritic spine of
postsynaptic neuron

FIGURE 5.9. NMDA-type glutamate receptors and AMPA-type glutamate receptors. The NMDA receptors are activated by high-frequency bursts but not by background activity, because the voltage-dependent blockade of the NMDA receptor's ion channel by magnesium ions (Mg^{2+}) is only relieved when the postsynaptic membrane is depolarized to a level positive to -50 millivolts. The AMPA receptors are activated by both background activity and high-frequencey bursts. Adapted from L. R. Squire and E. R. Kandel, *Memory: From Mind to Molecules* (Scientific American Library, New York, 1999). *Joan M. K. Tycko, illustrator.*

If this process is the trigger for LTP, then drugs that block the NMDA receptor should also block LTP. This is indeed what happens in most hippocampal synapses. Furthermore, if one injects neurons with drugs that rapidly bind calcium ions as soon as they enter the cell, thereby preventing them from interacting with other molecules, this will also prevent LTP. Calcium ions entering through NMDA receptors can activate lots of different calcium-sensitive enzymes in the neuronal dendrite. Rapid, large calcium transients can activate an enzyme called calcium/calmodulin protein kinase II alpha, typically abbreviated CaMKII. This enzyme transfers chemical phosphate groups onto proteins to change their function. Although the substrates of CaMKII action relevant for LTP are not known, one popular hypothesis is that this process ultimately results in the insertion of new AMPA-type receptors into the postsynaptic membrane, thereby strengthening the synapse. It should be mentioned that though this NMDA receptor → CaMKII → AMPA receptor insertion cascade is the most common form of LTP, it is not the only one. There are others that can use different biochemical steps and produce LTP through different means (such as increased glutamate release or increased conductance of existing AMPA receptors).

What about LTD? How does sustained synaptic activation at moderate frequencies result in persistent synaptic weakening? Interestingly, in its most common form, LTD also uses the NMDA receptor. In this case, moderate frequency stimulation results in partial relief of the magnesium ion blockade of the NMDA receptor. This produces a calcium flux, but one that is small and sustained rather than large and brief. Small, sustained calcium signals are insufficient to activate CaMKII and therefore don't produce LTP. Instead, they activate an enzyme that does the opposite job: protein phosphatase 1 (PP1) removes phosphate groups. Activation of PP1, not surprisingly, ultimately results in the removal of AMPA receptors from the postsynaptic membrane, thereby depressing synaptic strength in a way that is the functional opposite of LTP. This LTD cascade involving NMDA receptor → PP1 → AMPA receptor internalization is a dominant form of LTD in the hippocampus, but it is only one of several mechanisms for producing persistent depression of synaptic strength.

Thus both LTP and LTD can be produced in several different ways. In reality, some individual synapses are able to express multiple forms of both LTP and LTD.

So, the model being developed here is that somehow, memory for facts and events, including memory for the location of the escape platform in the Morris water maze, is encoded by producing LTP and LTD in an array of hippocampal synapses, and these forms of LTP and LTD are critically dependent upon triggering by NMDA receptors. One central test of this hypothesis was to inject rats with NMDA receptor–blocking drugs to see if they could learn the Morris water-maze task in conditions where LTP and LTD were mostly blocked. This experiment, which has now been repeated several times in different labs, showed that spatial memory was indeed severely impaired under these conditions. Later, a similar result was obtained by using mutant mice that failed to express functional NMDA receptors in a crucial subregion of the hippocampus (called area CA1; see Figure 5.10). In all of these cases, the general sensory and motor functions of these mice were largely intact—the failure in the maze task appeared to be a genuine memory deficit and not a trivial defect in seeing or swimming or stress level.

Would it be possible to train rats in the Morris water maze task and then analyze their hippocampal tissue? Many attempts have been made over the years to look for the electrical, biochemical, or structural correlates of learning in the hippocampus. There have been intermittent claims, but in truth not much has come of these efforts. Here's the problem. Spatial learning is likely to produce changes in a very small fraction of spatially distributed hippocampal synapses, and we don't have a good way to know where these synapses are. So, whether you're recording synaptic strength electrically or making biochemical or structural measurements, there's a big "needle-in-a-haystack" problem: it's almost

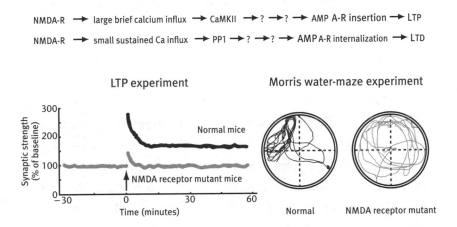

NMDA-R → large brief calcium influx → CaMKII → ? → ? → AMP A-R insertion → LTP

NMDA-R → small sustained Ca influx → PP1 → ? → ? → AMPA-R internalization → LTD

LTP experiment **Morris water-maze experiment**

Synaptic strength (% of baseline)

300

200 — Normal mice

100 — NMDA receptor mutant mice

0

−30 0 30 60

Time (minutes)

Normal NMDA receptor mutant

FIGURE 5.10. An experiment showing that mutant mice lacking functional NMDA-type glutamate receptors in a crucial region of the hippocampus have impaired LTP, LTD, and spatial learning. The top panel shows the signaling cascades triggered by the NMDA receptor to induce both LTP and LTD. The question marks indicate that there are multiple steps leading to AMPA receptor insertion and internalization that we still do not understand. The lower left panel is a plot of synaptic strength as a function of time in an LTP experiment. LTP was induced by applying high-frequency bursts to the presynaptic axons at the point indicated by the upward arrow. The lower right panel shows the path of well-trained mice in a Morris water maze. These are the results of a probe trial in which the platform is removed to see where the mouse will hunt for it. The normal mouse has a well-established memory for the correct platform location in the upper left quadrant while the LTP/LTD-lacking mutant mouse has little memory for the location and therefore searches widely in the water maze. Adapted from J. Z. Tsien, P. T. Huerta, and S. Tonegawa, The essential role of hippocampal CA1 NMDA receptor-dependent synaptic plasticity in spatial memory, *Cell* 87:1327–1338 (1996). *Joan M. K. Tycko, illustrator.*

impossible to measure the relevant changes when they are diluted in a sea of other synapses that are not a part of the memory trace.

The results showing that treatments that interfere with hippocampal NMDA receptor function block spatial learning in rats and mice *suggest* that our working model is correct: the engram for declarative memory in the hippocampus requires LTP and LTD. Do these findings *prove* this hypothesis? Unfortunately, no. Although hippocampal NMDA receptor manipulations interfere with spatial learning, attempts to interfere with LTP and LTD by targeting biochemical signals that follow NMDA receptor activation have met with mixed success. One can block most forms of LTP or LTD by interfering with CaMKII or PP1 or certain types of AMPA receptors, but this will not always produce a deficit in spatial learning tasks. In addition, it's very likely that these manipulations affect many processes in addition to LTP and LTD. The calcium flux through the NMDA receptor activates many enzymes, not just PP1 and CaMKII. There is further divergence of the signaling cascade as we move along: CaMKII, for example, transfers phosphate groups to hundreds of proteins in hippocampal neurons, not just those involved in LTP. As a consequence, one cannot be completely confident that the blockade of spatial learning produced by these drugs or molecular genetic tricks is really due to an LTP/LTD deficit as opposed to some side effect.

To summarize, we know that destroying the hippocampus will prevent spatial learning in rats and mice, and there is suggestive, but not conclusive, evidence that memory for locations in space is stored in the hippocampus by changing the strength of synapses through LTP and LTD. How does making certain synapses weaker or stronger in the hippocampus give rise to the behavioral memory that allows an animal to learn the Morris water maze or another spatial task? The short answer is that we don't know. The hippocampus is not anatomically or functionally organized in a way that makes this obvious. The slightly longer answer is that even through we don't know, there is an interesting hint that may be relevant to this difficult problem.

John O'Keefe, Lynn Nadel, and their coworkers at University College, London, made recordings from neurons in the hippocampus of rats as they explored an artificial environment in the lab. What they found was that about 30 percent of one class of cells in the hippocampus (called pyramidal cells) seemed to encode the animals' position in space. When a rat is placed in a new environment and has a chance to explore, recordings will reveal that, after a few minutes, one cell fires only when the animal is in a particular location (say, the upper left edge of a large circular cage (see Figure 5.11). This particular cell, referred to as a "place cell," will once again fire in this fashion even if the rat is removed from this environment and returned days to weeks later. Place cells have been found in mice as well as rats. Recording from additional cells reveals that there are place cells that fire specifically for all different parts of the explored environment. Some are fairly sharply tuned for place (Figure 5.11) while others fire over a broader area. When place cells are recorded in mutant mice in which the hippocampal pyramidal cells have been engineered to have a form of CaMKII that is always on (they can't have more LTP because it is already turned up to maximum levels), interesting properties emerge. Place cells do form characteristic firing patterns when the mouse explores an environment, but then, when the animal returns to the environment, the tuning of these cells tends to change (Figure 5.11). Because these mutant mice are also impaired in spatial learning tasks, it has been suggested that LTP is required to maintain the tuning of place cells and that these place cells form a cognitive map of space that allows the animal to store spatial memory.

The problem is that the physical details of this cognitive map in the hippo-

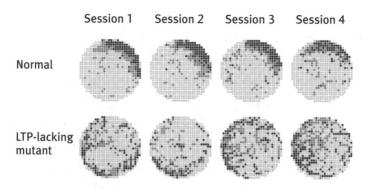

Session 1   Session 2   Session 3   Session 4

Normal

LTP-lacking
mutant

FIGURE 5.11. Place cells in the hippocampus of mice. These figures show the firing
rate of individual pyramidal cells in the hippocampus of a mouse explor-
ing a circular environment. Black pixels indicate high firing rates, and
light gray low firing rates, for that particular location. Place cells from
normal mice may be either sharply tuned (as shown here in the example
from the normal mouse) or broadly tuned, but the representation of
space is stable over many training sessions. Recordings made from a mutant mouse that lacks LTP (it has been engineered to have a form of
CaMKII that is always turned on in these neurons) show that place
fields are not stable over repeated training sessions. This correlates with a
failure of these mutant mice to learn certain spatial tasks. Adapted from
A. Rotenberg, M. Mayford, R. D. Hawkins, E. R. Kandel, and R. U.
Muller, Mice expressing activated CaMKII lack low frequency LTP and
do not form stable place cells in the CA1 region of the hippocampus,
Cell 87:1351–1361 (1996). *Joan M. K. Tycko, illustrator.*

campus are not obvious. Sensory systems have maps that represent the external
world anatomically in the brain: adjacent cells in the primary visual cortex will
be activated by light coming from adjacent points in the visual field. Likewise,
adjacent cells in the primary somatosensory cortex will be stimulated by touch
on adjacent points on the body surface. But although different hippocampal
neurons that code for the same location in space tend to fire together, they are

not physically organized in any coherent fashion. One cell that codes for the upper left quadrant of the environment may be located at the opposite end of the hippocampus from another cell that codes for the same region, and cells in the intervening tissue are not organized in any fashion to represent the spatial world. So, while we are beginning to gain an understanding of the molecular processes that represent experience as changes in neuronal function (LTP, LTD, and changes in intrinsic excitability) and structure, and there is some evidence emerging to link these processes with specific forms of learning, we are still far, far away from a complete "molecules-to-behavior" explanation of declarative memory.

WE HAVE SEEN that the brain does not use a single cellular process or a single brain region to store memory. Rather, memory storage involves multiple brain locations and several broad classes of mechanism (synaptic plasticity, intrinsic plasticity), each of which can be produced by a number of different molecular strategies. Crucially, the cellular and molecular mechanisms of memory storage are not unique. In true, kludgy evolutionary fashion, the mechanisms for storing memory have been largely adapted from those designed to wire up the brain in response to experience during the later stages of development (during late pregnancy and early childhood).

Let's put this back in a historical context. The design of the brain has been limited as it has evolved by three main considerations.

1. During the course of evolution, the brain has never been redesigned from the ground up. It can only add new systems onto existing ones.
2. The brain has a very limited capacity for turning off control systems, even when these systems are counterproductive in a given situation.

3. Neurons, the basic processors of the brain, are slow and unreliable, and they have a rather limited signaling range.

These considerations have driven the brain's solution to the problem of building computational complexity: a brain that has a huge number of neurons and in which these neurons are highly interconnected. This big complex brain creates two problems. How do you get a large head through the birth canal? And how do you specify the wiring diagram for 500 trillion synapses genetically? The solutions, as previously discussed, have been to only roughly specify the wiring diagram of the brain genetically and to reserve significant brain growth and synapse formation until after birth. This design allows a head that can pass through the birth canal. It also allows sensory experience to guide the fine-scale wiring of the brain. In order to do that, there had to be mechanisms by which particular patterns of sensory experience could drive changes in synaptic strength (LTP and LTD), intrinsic excitability, and the growth and retraction of axonal and dendritic branches as well as synapses. These, of course, are the same cellular and molecular mechanisms that, with slight elaboration, are retained in the mature brain to store memory.

This is the ultimate example of "when life gives you lemons, make lemonade." Our memory, which is the substrate of our consciousness and individuality, is nothing more than the accidental product of a work-around solution to a set of early evolutionary constraints. Put another way, our very humanness is the product of accidental design, constrained by evolution.

Chapter Six

Love and Sex

HUMANS ARE TRULY the all-time twisted sex deviants of the mammalian world. I'm not saying this because some of us get turned on by the sight of automobile exhaust systems, the smell of unwashed feet, or the idea of traffic cops in bondage. After all, other species are at a disadvantage in expressing their kinks by not having reliable access to the Internet. Rather, I mean that the more prosaic aspects of sexual activity in humans are far outside the mainstream of behavior for most of our closest animal relatives.

The spectrum of human amorous and sexual behavior is wide and deeply influenced by culture (and I will consider these issues shortly), but let's first talk about the generic presumed norm: regular, old-fashioned monogamous heterosexual practice. Then we can see how it compares with the practices of most other mammals. The simplified human story, stripped of all the romance, is something like this. Once upon a time, a man and woman met and felt mutual attraction that they codified in a ceremony (marriage). They liked privacy for their sexual acts and they declined opportunities for sex with others. They had sex, including intercourse, many times, in most phases of the woman's ovulatory cycle, until she became pregnant. Once it was known that the woman was pregnant, they continued to have sexual intercourse for some time thereafter. After the baby was born, the man helped the woman to provide resources and sometimes care for the child (and for the other children that followed). The woman and man continued their monogamous relationship and remained sexually active well beyond the woman's childbearing years, as marked by her menopause.

Now let's hear another perspective. The comedian Margaret Cho uses the line "Monogamy is sooo weird . . . like . . . when you know their name and stuff?" This brings down the house in a comedy club, but the idea is actually the dominant one in the nonhuman world: more than 95 percent of mammalian species do not form lasting pair bonds, or even pair bonds of any kind. In fact, rampant sexual promiscuity is the norm for both males and females, and this promiscuous sex is typically conducted in the open, for everyone in the social group to see. One-night stands and public sex are the rule, not the exception. One consequence of all this public promiscuity is that in most nonhuman mammals the father makes little or no contribution to rearing the young. In some cases, the male does not stay in a social group following mating, but rather drifts away. In others, the male stays in the social group but does not appear to recognize his own offspring.

This arrangement may give the impression that most nonhuman animals are libertines, but in another sense they are deeply conservative. Humans often have sex when it is either unlikely or impossible for conception to occur (during the wrong part of the ovulatory cycle, during pregnancy or after menopause), but most nonhuman mammals have sex that is very accurately timed to match ovulation. Human females have concealed ovulation: it is almost impossible for a male to detect directly the female's most fertile days. Although women are able to train themselves to detect ovulation, there is no evidence of an instinctive knowledge of ovulation like that possessed by other female primates. In fact, while many studies have been done on this topic, it is not clear that women are most interested in sexual intercourse during the preovulatory (fertile) phase of their cycle.

In contrast, most nonhuman females in the mammalian world advertise their impending ovulation with sexual swellings, specific odors, or stereotyped sounds and gestures (such as a posture that presents the genitals) indicating sexual interest. Typically, neither males nor females will approach each other for sex during nonfertile times. Sex after menopause is not an issue because although nonhuman females do show gradually declining fertility after a certain age, there is no point where they become absolutely infertile. Indeed, menopause may be a uniquely human phenomenon.

Of course, these human sexual distinctions are based on a broad generalization. There are some nonhuman species such as gibbons and prairie voles that form long-term pair bonds in which the father helps rear the young. There are also a few animals, such as dolphins and bonobos, that seem to share the human proclivity for recreational sex, and some others, such as vervet monkeys and orangutans, where the females have concealed ovulation. On the human side, it is not all Ozzie-and-Harriet either: clearly, humans are not all monogamous (or even serially monogamous), and in some cultures or subgroups polygyny (multiple wives) or polyandry (multiple husbands) is an established practice. Nonetheless, it is clear that the dominant human practice, across cultures, is monogamy, or at least serial monogamy. The critical point here is that in humans, most females have a single sexual partner in a given ovulatory cycle. In studies where paternity has been evaluated with genetic tests across large numbers of children, the vast majority (over 90 percent) of children are indeed found to be the offspring of the mother's husband or long-term partner, and most fathers provide some form of care and support for their children (although this may take the form of providing food, protection from others, shelter, and money rather than direct child care).

We share a number of common sexual practices with other animals. Oral-genital stimulation (of both sexes) is one of these. Masturbation is another. Both male and female animals have been observed to masturbate, some even using objects to do so. To date, however, humans are the only species reported to masturbate while watching *Richard Simmons' Sweatin' to the Oldies, Disc 2*. Originally it was thought that masturbation might solely be a phenomenon of animals in captivity, but there are now reliable field reports of both male and female masturbation in wild bonobos and red colobus monkeys. There is also evidence for nonhuman masturbation independent of direct genital stimulation. Sir Frank Darling, in his classic 1937 book on animal behavior *A Herd of Red Deer*, reported that during the rutting season male Scottish red deer masturbate "by lowering the head and gently drawing the tips of the antlers to and fro through the herbage." This typically results in penile erection and ejaculation within a few minutes. Finally, it should be mentioned that both male and female homosexual acts have been observed in a large number of mammalian species, although, to my knowledge, there are no reports of lasting homosexual pair bonds in nonhumans.

So, why have humans evolved such a distinct cluster of sexual behaviors with concealed ovulation, recreational sex, long-term pair bonding, and prolonged paternal involvement? Though a few of our close simian cousins share some of these traits—the bonobos with their penchant for recreational sex and gibbons with their long-term pair bonding—none of these species has the complete cluster of behaviors. Thus these aspects of human sexual behavior are likely to be recent evolutionary developments in our primate lineage.

What I will argue here is that our normative human sexual practices follow directly from inelegant brain design. Let's work backward to try to explore this question. Why do humans have concealed ovulation and recreational sex? One persuasive evolutionary hypothesis, from Katherine Noonan and Richard Alexander of the University of Michigan, is that concealed ovulation functions to keep the male around. Let's first consider the counter-example: When ovulation is clearly advertised, the male can maximize his reproductive success by mating with a given female in her fertile time and then, when her fertile time is over, leaving to try to find a another fertile female to impregnate. In this system, the male does not have to worry that some other male will come along and impregnate the first female while he is away because he knows that she is no longer fertile. This is the mating system found in many species, including baboons and geese. With concealed ovulation, however, the couple has to mate all through the woman's cycle to have a reasonable chance of conceiving. Not only that, but if the male decides to stray and try his luck with another female, he cannot be sure that another male will not sneak in the back door and mate with the first female on her fertile days. Furthermore, his chance of finding another ovulating female is low. Hence, with concealed ovulation, the best male strategy is to stick with one female and mate with her all the time.

Enough about the male. What does the female get out of this arrangement? Isn't her best reproductive strategy to play the field in the hope of getting the best-quality male genetic contribution to her offspring? Indeed, the females of many species, including many mammalian species, do exactly that. The crucial difference is that although a female orangutan, for example, easily rears her offspring alone, human females don't have it so easy. Most other animals are able to find their own food immediately after weaning, but human children do not achieve this level of independence for many more years. As a consequence, the reproductive success of a female human is much greater if she can establish a long-term pair bond with a male and he contributes in some form to child-rearing. Males tend to buy into this arrangement for two reasons. One is that if the male plays along he can be confident of paternity: he won't be wasting his resources supporting the offspring of another male. Another is that he, and the female, will enjoy the bonding that comes from frequent sex. This bonding and reward is enough to keep humans having sex even when conception is impossible (during pregnancy or after menopause).

In this story, the key point is that human females need male help in certain aspects of childrearing much more than females of other species because human infants are totally helpless and even toddlers and small children are incapable of fending for themselves. Why is that? Recall that the human brain at birth has only about one third of its mature volume and that early life is crucial for the experience-dependent wiring and growth of the brain. The human brain grows at an explosive rate until age 5 and it is not completely mature until about age 20. Unlike the 5-year-olds of most other species, human 5-year-olds simply do not have sufficiently mature brains to find their own food and protect themselves from predators.

Let's summarize by telling the story back in the other direction. Human brains are never designed from the ground up. Rather, as we have seen, new systems are just added on top of the evolutionarily older ones below. This means that the brain must grow in size as it evolves new features. Even more important, the brain is made of neurons that haven't changed substantially in their design since the days of prehistoric jellyfish: as a consequence, neurons are slow, leaky, unreliable, and have a severely limited signaling range. So, the way to build sophisticated computation in a brain with these suboptimal parts has been to create an enormous, massively interconnected network of 100 billion neurons and 500 trillion synapses. This network is too big to have its point-to-point wiring diagram explicitly encoded in the genome, so experience-driven "use it or lose it" rules for wiring must come into play to actively construct this huge network. This necessitates extensive sensory activity, which mostly proceeds after birth, and this requires an unusually protracted childhood during which the brain matures. In addition, the physical constraints of the birth canal make it impossible for a human baby to be born with a more mature brain—it just wouldn't fit. As it is, death during childbirth is a significant human phenomenon, particularly in traditional societies, whereas it is almost unknown among our closest primate relatives.

As a consequence of all this, human females are uniquely dependent on male support to raise their offspring. They secure their reproductive success by having concealed ovulation, which compels males to adopt a strategy of mating with one female repeatedly throughout her cycle. This monogamous, mostly recreational sex has two effects: it gives a high probability of accurately knowing paternity of the resultant offspring and it helps to reinforce a lasting pair bond, both of which promote continued care of the offspring by both parents. Or, to reduce it to an extreme level of speculation: if human neurons were much more efficient processors, then heterosexual marriage might not exist as a dominant cross-cultural human institution.

"But . . . but . . . but," I can hear you say, "does this explanation really bear on how we live now? After all, in my city there are plenty of single moms who are raising their children just fine, thank you. And there are couples lining up to adopt children with whom they share no genetic material and others who are happy not to have children at all. There are plenty of gay people, and a few are raising children, but most are not. Also, there are lots of people who are having sex outside of their long-term relationship. Though this is all true, several points, central to how human sexual behavior has developed, must be made in thinking about these observations. First, evolution is a slow process and our genomes are never totally adapted to rapidly changing conditions. In our modern world, some very recent changes relevant to sexual behavior, such as the availability of contraception and assisted fertility, and changes in social conventions, political systems, and technologies, have allowed women to live independently. Most of these changes have only appeared in the last generation. So, the genes that help to instruct the parts of our brains involved in sexual behavior have not yet undergone selection by many of the forces operating in modern society. Indeed, this is a general issue in considering human evolution that applies to many aspects of biological function, not just the biological basis of sexual behavior. Second, certain drives related to sexual function will persist even in situations or stages of life where they are no longer relevant to getting one's genes into the next generation. Hence, people routinely feel sexual attraction and even form long-term pair bonds (read as "fall in love") in situations where there is no chance of producing offspring (because of contraception, infertility, menopause, same-sex partner, and so on). Likewise, many couples feel a strong urge to raise children, even if those children do not share their genes. Third, even in our modern society with sex outside of long-term relationships, high divorce rates, and so forth, it is amazing that the end result of these factors on paternity is minor. As I mentioned previously, widespread genetic tests of paternity across several cultures have shown that over 90 percent of children were indeed fathered by the mother's husband or long-term partner. Furthermore, despite divorce and remarriage, a similar fraction of fathers contribute in some fashion to raising their children. In this way, though cultural attitudes and practices may differ, the ultimate outcome of the sexual lives of today's New

Yorkers or Londoners is not very different from that of people living in more traditional societies.

I'VE SPENT A good bit of time describing a rather limited spectrum of human sexual behavior and speculating about how inefficient brain design might have helped to create it. Let's now turn our attention to the other side of the coin: how brain function influences our sexual and amorous drives. In doing so, we must first consider the prerequisite of all sexual behavior, which is the development of gender identity. How do we come to see ourselves as male or female?

Gender identity is a complex process in which biological and sociocultural factors come together. It's not sufficient to say that if your sex chromosomes are XX and you have ovaries and a vagina you will then think of yourself as female, whereas XY chromosomes, testicles, and a penis will cause you to think of yourself as male. It's more complex in at least two ways. As you undoubtedly know, there is a small fraction of people who feel gender dysphoria. These people believe deeply that their chromosomal sex does not match their sense of self. This, despite the outward characteristics of their bodies and overwhelming social pressure. In some more affluent cultures, these transgendered people will often elect to cross-dress, take hormonal treatments, or undergo various forms of surgery to partially or completely reassign their sex. Gender dysphoria is more common in chromosomally male people, but it is not solely a male-to-female phenomenon. It's worth noting that while gender dysphorics, if allowed by social convention, will almost always cross-dress, the reverse isn't true: most cross-dressers identify with their chromosomal sex and do not experience gender dysphoria. Rather, they cross-dress as a more subtle expression of sexual identity.

Once you have self-identified as male or female, what this means in terms of your ideas and expectations is hugely influenced by culture and personal experience. The idea of what it means to be a man or a woman varies widely across cultures, families, and even individuals, in ways we know all too well: Japanese female gender identity, for example, is not the same as Italian female gender identity. In recent years, our cultural ideas about male and female identity have undergone rapid change. Perhaps the clearest examples of culturally constructed gender identity may be found in those traditions that have institutionalized transgendered status. In many Native North American cultural groups, a practice called two-spirit flourished. In these traditions, chromosomal males who identified as females and, to a lesser extent, chromosomal females who identified as males were encouraged to cross-dress and were accorded a special shamanistic status for their abilities to bridge the male and female worlds. In Polynesia, there is a tradition in which the first-born child is designated as a sort of mother's helper and assigned a female-typical social role. In some cases this is done irrespective of the chromosomal sex of the child, and the male-to-female transgendered people who result are given the name *mahu* (in Tahiti or Hawaii) or *fa' a fafine* (in Samoa). An early European encounter with this practice was recorded by a Lieutenant Morrison, a member of Captain William Bligh's 1789 expedition to Tahiti: "They have a set of men called mahu. These men are in some respects like the eunuchs of India but they are not castrated. They never cohabit with women but live as they do. They pick their beards out and dress as women, dance and sing with them and are as effeminate in their voice. They are generally excellent hands at making and painting of cloth, making mats and every other woman's employment." Like Native American two-spirit practitioners, mahu are assigned a high social status and are considered both lucky and powerful. King Kamehameha I of Hawaii made sure to have mahu dwelling within his compound for exactly this reason. The larger issue here, as illustrated by mahu, two-spirit, or merely your hyper-macho Uncle Fergus, is that although sex is simply determined by sex chromosomes and the resultant action of sex hormones, gender identity is a more complex process in which there is an interplay of biological and sociocultural factors.

Can we identify differences in male and female brains that might underlie the biological component of gender identity? Male brains, on the average, are slightly bigger than female brains, even when a correction is made for body size. This is most apparent in measures of the thickness of the right cerebral cortex. More interestingly, a particular cluster of cells in the hypothalamus called INAH3 (an acronym for interstitial nucleus of the anterior hypothalamus number 3) is two to three times larger in men than in women. This is very suggestive because the cells of INAH3 have an unusually high density of receptors for testosterone and also because neural activity in this region is correlated with certain phases of male-typical behavior during sex (more on this later). Lest we start to think that everything is bigger in males, there are two key regions that are proportionally larger in the female brain. These are the corpus callosum and the anterior commissure. These structures are bundles of axons (white matter) that carry information from one side of the brain to the other. They are particularly important in linking the two sides of the highest and most recently evolved brain region, the cerebral cortex. It is almost certain that this list is incomplete in several respects. There are likely to be more regional size differences in male versus female brains that will emerge with further research. In addition, there are likely to be even more differences that will be manifest not as size differences but rather as differences in cellular structure (such as degree of dendritic branching), biochemical constituents (perhaps the density of neurotransmitter receptors or voltage-gated ion channels), or electrical function (such as spiking rate and timing in particular neurons).

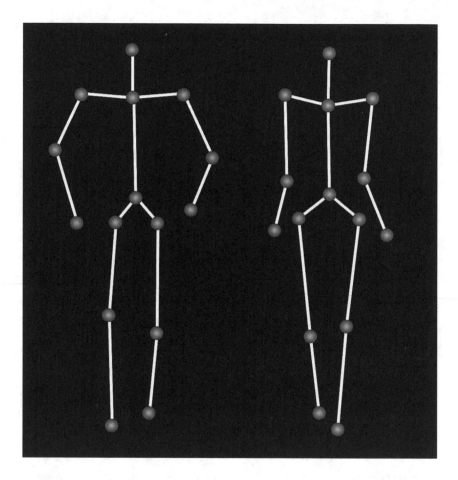

FIGURE 6.1. ¿Qué es más macho? Even though there is very little information in these stick figures, we can easily assign them to typically male and female categories. This illustrates that our brains' visual systems have become very highly specialized for gender recognition. *This image was kindly provided by Professor Nikolaus Troje of Queens University, Ontario, Canada. The male/female distinction is even more apparent when the figures are animated with gender-typical walking motions. You can see this at Professor Troje's website: www.biomotionlab.ca/Demos/BMLgender.html*

In addition to these neuroanatomical differences between men and women there are some consistent behavioral differences. This has been a contentious and politically charged area of research, but a large number of studies conducted by different groups around the world now seem to be pointing to a consistent set of conclusions. On average, women score better than men on some language tasks, such as rapidly generating words in a particular category. This is called "verbal fluency" and has been found cross-culturally. They outscore men in tests of social intelligence, empathy, and cooperation. On average, women are better at tasks that involve generating novel ideas, and they excel at matching items (spotting when two items are alike) and arithmetic calculation. But men generally outperform women on tests of mathematical reasoning, particularly those using word problems or geometry. They are better at some spatial tasks such as mental rotation of three-dimensional objects and distinguishing figures from the background. The general conclusion is that, on average, women and men do tend to have different cognitive styles. Of course, these differences are seen in averages of large populations and individual men and women can have abilities throughout the performance range for all of these traits. Tests that seek to measure general intelligence have not found significant differences between large male and female populations.

So, we have some evidence for differences in male and female brain structure and some differences in male and female mental function. One key issue is to what degree these anatomical and behavioral differences are genetically versus socioculturally determined, the old nature-versus-nurture question. The fact that we can see anatomical differences between adult male and female brains does not, in itself, prove that these differences have a genetic basis. Recall from Chapter 3 that experience can mold neuronal connections and fine structure as particular patterns of electrical activity give rise to expression for certain genes. Perhaps the way a typical girl is raised causes her to grow a somewhat larger set of axonal connections between the left and right sides of the brain (the anterior commissure and corpus callosum) and the way an average boy is raised causes expansion of INAH3.

At present, although it seems reasonable to imagine that sociocultural factors might affect sex differences in brain structure, there is no evidence leading to either acceptance or rejection of this idea. But several lines of evidence argue for a genetically based explanation. For example, accumulating evidence indicates that gender differences in behavior can be seen very early in life and across species. On average, newborn girls spend more time attending to social stimuli such as voices and faces while newborn boys show greater fascination with spatial stimuli such as mobiles. Young male monkeys and rats tend to engage in more rough-and-tumble play than their female counterparts. Young male rats perform better in spatial maze tasks than females.

Correlational studies on both girls and boys have shown that the levels of prenatal testosterone can predict performance on some spatial tasks when it is measured later in life. Although testosterone is thought of as a "male hormone" deriving from the testes, it is also produced in the adrenal glands and is therefore present in females in smaller amounts. In one recent report, Simon Baron-Cohen and his colleagues at the University of Cambridge found that children exposed to a high level of testosterone in utero were less likely to make eye contact at the age of 12 months and had less developed language skills at the age of 18 months. In sum, testosterone exposure seems to drive a more male-typical cognitive/behavioral style even when this is measured quite early in life.

Extreme examples of this idea can be found in cases where sex hormones are subject to unusual manipulations. Girls suffering from a swelling of the adrenal glands called congenital adrenal hyperplasia, or whose mothers were treated during pregnancy with the steroid drug diethylstilbestrol (DES), are exposed to much higher than usual levels of testosterone, starting in utero. On average, these girls tend to perform more like boys in some cognitive tests (better at mathematical reasoning and spatial tasks). Their behavior as small children was also more boylike: these girls displayed more aggressive play and showed more interest in object toys (trucks) than social toys (dolls). An analogous result is seen in animal experiments: the performance of a group of female rats in a spatial maze task was increased to average male levels when they were treated with testosterone shortly after birth.

The converse finding is revealed in a disease called androgen-insensitivity syndrome, in which males develop normal testes that secrete typical levels of testosterone, but a mutation in the receptor for testosterone (a "male hormone," or androgen) renders cells unable to respond to these compounds and so the body (and the brain) develop on a female path. When visual-spatial ability was studied in such a population by Juliane Imperato-McGinly and her coworkers at Cornell Medical School, it was found that not only did they perform more poorly than average males, but they did significantly worse than average females as well. Presumably this reflects the fact that males with androgen insensitivity syndrome receive no effects of androgens at all during development and early life, while normal females do have a low level of androgen exposure from adrenal gland–derived testosterone. These results are similar to those seen in male rats castrated at birth, which also perform more poorly than females in a spatial maze task.

Some sex-based differences in cognitive style may be attributed to differences in brain structure. Melissa Hines and her coworkers, then at UCLA, found that in a population of normal women, those with the largest corpus callosum, in particular, a subregion of the corpus callosum called the splenium, performed best on tests of verbal fluency. They hypothesized that a larger splenium allows a greater flow of information between language centers in the left and right brains.

THE ISSUE OF sex-based differences in brain function and cognition entered the public consciousness in a big way when Larry Summers, president of Harvard University, addressed the National Bureau of Economic Research Conference on Diversifying the Science and Engineering Workforce on January 14, 2005. He proposed that the extreme underrepresentation of women at the highest levels of science and engineering could be partially explained by genetic differences underlying the function of male and female brains, what he called "different availability of aptitude at the high end." He proposed that if one were to evaluate the top 2 percent or so of scorers on standardized math or science tests, one would find four times more men than women; and he suggested that this difference in the elite pool underlies, in part, the underrepresentation of women in science and engineering, particularly at top-ranked universities. These comments evoked a firestorm of criticism and counterattack that continued even after his resignation some months later.

Let's evaluate the Summers hypothesis in light of the work I've just reviewed on sex-based differences in brain structure and cognitive style. There is reason to believe his foundational premise: one can devise cognitive tests that reveal differences between men and women in both the average score and in the variation of scores (greater variation will influence the top 2 percent). But, crucially, are these tests predictive of success in science and engineering and the very top ranks? To my knowledge, data are not available to assess this point. But from my own experience in science, I would guess not. I have had the pleasure of interacting with many of the world's most successful scientists (albeit, mostly biologists) over the years and one thing is clear: there is not a single cognitive strategy that underlies success at the very top of science. Some of the world's top scientists think in terms of equations, others verbally, others spatially. Some rely on step-by-step deduction and logic to reach their conclusions while others have a flash of insight that they must then go back and test post hoc to see if it is valid. Einstein, by all reports, was a rather middling mathematician, but this did not stop him from making paradigm-shifting contributions to physics that were expressed mathematically.

For the Summers hypothesis to be true, the cognitive differences that one measures on standardized tests must truly be predictive of scientific success at the high end. In addition, for the Summers hypothesis to be true, a diminished pool of elite women must be a limiting factor in the development of elite scientists. Here, my personal experience also leads me to be skeptical. I served as Admissions Chair and then Director of the Graduate Program in Neuroscience at The Johns Hopkins University School of Medicine from 1995 to 2006. That is one of the top programs in the world and it draws a very talented pool of students. During this period, comparable numbers of men and women enrolled. Similar numbers of men and women completed the program and those that did had similar productivity (measured, for example, as numbers of papers in the most prestigious journals). But as these students moved on in their careers, the women began to drop out. Fewer entered postdoctoral fellowships. Of those who finished their fellowships, fewer applied for faculty positions at elite universities, and of those that applied, fewer were successful in obtaining the top positions. This is reflected in the composition of my own department, where only 3 of 24 faculty are women. At least in the field of neuroscience, I highly doubt the validity of the Summers hypothesis: there are plenty of women with the very highest scientific aptitude in the pool, but the pipeline is leaking prodigiously, for a variety of reasons. These reasons include many social factors including a hostile environment, inflexible tenure and promotion policies (that do not account for a woman's childbearing years), and, in some cases, blatant discrimination.

The Summers episode has been damaging to science for at least two reasons. First, undoubtedly, some women who might otherwise have considered a career in science and engineering have reconsidered in light of his remarks, either because they accepted his hypothesis or because they took his remarks as evidence of a hostile environment to women scientists in academia. Second, the backlash against Summers has included statements indicating that even considering the issue of sex-based differences in brain function or cognitive style should not be allowed. It's easy to see where these "politically correct" ideas come from. It's natural to be suspicious of work that could be used to rationalize the status quo, in this case male-dominated science. But this position is fundamentally intellectually dishonest. At some level, genes and epigenetic factors, including those linked to sex, influence the cognitive style of male and female populations. Pretending that this isn't true does not advance the cause of the rights of women (or any other historically oppressed group, for that matter). Scientists should be able to advocate for a woman-friendly, merit-based, inclusive, and diverse scientific enterprise without denying the mounting evidence for sex-based differences in brain function and cognitive style.

ENOUGH ABOUT MALE and female brains. Let's talk about love and sex. The 1970s art-rock band Roxy Music summed it up rather succinctly when they sang, "Love is the drug got a hook in me." What is the neurobiological basis of this? Is love, or at least sex, really like a drug? Not surprisingly, we know more about the brain's involvement in sexual acts than we do about its role in love and attraction.

Andreas Bartels and Semir Zeki of University College, London, have taken

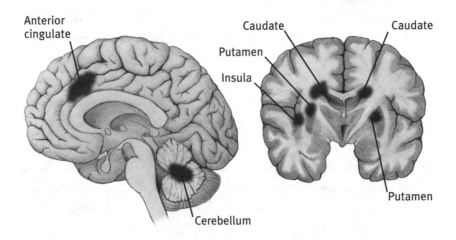

Anterior cingulate

Caudate

Caudate

Putamen

Insula

Cerebellum

Putamen

FIGURE 6.2. Specific brain activation produced by viewing photos of a lover's face. Left: A slice view down the midline of the brain with the nose oriented to the left created by a brain scanner. Right: A different slice view, crosswise, just in front of the ears. The black patches in both panels show the regions of activation. Adapted from A. Bartels and S. Zeki, The neural basis of romantic love, *Neuroreport* 11:3829–3834 (2000). *Joan M. K. Tycko, illustrator.*

an interesting approach to finding neural correlates of romantic love. They recruited male and female subjects in their 20s who claimed to be "truly, deeply, and madly in love" and imaged their brains while they looked at photographs of their lovers' faces. Then, they performed a similar experiment using photographs of friends for whom the subjects had no strong amorous or sexual feelings, matched for age, sex, and duration of friendship with the idea that the former minus the latter would reveal sites of brain activation specific to romantic love, as opposed to merely vision or face recognition. This calculation found increased activity in several discrete locations including the insula and anterior cingulate cortex (areas known to be important in processing emotional stimuli) when the subjects were viewing the lover's face and, surprisingly, in two regions that are mostly known for their involvement in coordination of sensation and movement: the caudate/putamen and the cerebellum (Figure 6.2). There was also a group of regions where activation was decreased with viewing the lover's face, and these included several regions of the cerebral cortex as well as the amygdala (an emotion, aggression, and fear center).

The Bartels and Zeki study used people who had been involved in a relationship for over 2 years. When this study was repeated by another group, headed by Lucy Brown at Albert Einstein College of Medicine, they recruited a subject pool of people who were in an earlier phase of their love relationships: ranging from 2 to 17 months. This population generally had the same pattern of activation as the longer-relationship subjects, with one consistent difference: the new relationship subjects also showed strong activation in the ventral tegmental area. This is particularly interesting because the ventral tegmental area is a reward center of the brain that is responsible for intensely pleasurable sensations. It is one of the key regions activated by heroin or cocaine. Like users of heroin or cocaine, new lovers frequently show very poor judgment, particularly about the object of their affections. So the boys in Roxy Music may have had it at least partially right. Love is a very powerful drug, but it only works for a while, a few months to a year—kind of like crack. Then the bloom falls off the rose, so to speak. Hence the old joke:

Q: Is it true you married your wife for her looks?
A: Yeah, but not the looks she's been giving me lately.

What can we take from all this? First, the caveats. This type of study is limited in several ways. It's correlational so we really don't know if any of these changes in brain activity are actually involved in the feeling of romantic love.

Also, it's a difficult study to perform. We don't really know the mental state of each subject as he or she gazed at their lover's photo, and it's difficult to completely exclude other factors in the results. For example, can the experimenters be confident that the lovers' faces were not simply more familiar to the subjects than those of their friends? But if we assume for a moment that the pattern of activation seen in this study does indeed reflect the activity of the brain during the feeling of romantic love, then one thing that's clear, and not surprising, is that not just a single, discrete region is involved. The fact that emotional and reward centers are activated is interesting. But what's surprising is the activation of centers for sensory-motor integration (caudate/putamen and cerebellum), which might shed new light on the issue.

Clearly, if you imagine a group of twenty-somethings looking at photos of their true loves you can guess that they might be feeling sexually aroused. How do the patterns of activation from the lover's face experiment compare to brain activation produced by seeing images of sexual activity? There is a collection of several studies of men and women who had their brain activity scanned while watching videos of strangers engaged in (hetero)sexual activity. In some of these studies the subjects were also asked to rate their level of sexual arousal in a questionnaire, and in one, performed by Bruce Arnow and his coworkers at Stanford, males had their arousal measured by means of a "custom-built pneumatic pressure cuff" attached to the penis with a condom. To try to isolate brain activation that was specific to sexual arousal, scans of these males were compared to scans of the same subjects watching presumably sexually neutral material such as landscapes or sports. The patterns of activation with the sex videos (Figure 6.3) was somewhat different in studies from various labs, but, in general, the studies showed partial overlap with the brain regions activated by photos of lovers' faces (Figure 6.2). Both activated the anterior cingulate, insula, and caudate/putamen. In addition, the sex videos produced activation of visual as-

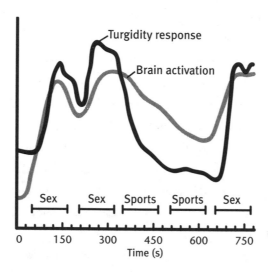

FIGURE 6.3. Simultaneous measurement of penile erection and brain activity while the subject watched alternating video snippets of sex and sports. Brain activity was recorded in a region called the insula. You can see that these two measures correlate rather well. Adapted from B. A. Arnow, J. E. Desmond, L. L. Banner, G. H. Glover, A. Solomon, M. L. Polan, T. F. Lue, and S. W. Atlas, Brain activation and sexual arousal in healthy, heterosexual males, *Brain* 125:1014–1023 (2002). *Joan M. K. Tycko, illustrator.*

sociation areas in the occipital and temporal cortex as well as some areas implicated in executive function and judgment in the frontal cortex. The sex videos did not activate the ventral tegmental reward area. Interestingly, Sherif Karama and his colleagues at the University of Montréal, who used both men and women in their study, found that only the men had significant activation in the hypothalamus. This result should be interpreted with caution, however, because it may reflect a difference in acculturated male and female responses to sex videos rather than an underlying difference in brain function.

Although it is difficult to derive a good understanding of either romantic love or the initial stages of sexual arousal from these human imaging studies, things become somewhat clearer as we begin to consider aspects of the sexual acts themselves, because this allows for animal experimentation: We can't easily ask animals how they are feeling, but we certainly can watch them have intercourse. Recall that unlike humans, most animals, including the rats and monkeys that are the mainstay of laboratory research, will only mate during the female's ovulatory phase. Therefore, the initiation of sexual behavior is typically under control by the hormonal processes of the female's ovulatory cycle. A female monkey comes into heat by a two-step process in which surges of the ovarian hormones estrogen and then progesterone occur. This process causes several different effects that prime sexual behavior. The estrogen acts, over a day or so, to stimulate the growth of synaptic connections in a region of the hypothalamus called the ventromedial nucleus. You may recall from Chapter 1 that this nucleus is also involved in feeding behavior—probably the ventromedial nucleus has different subdivisions devoted to eating and sexual behavior. Estrogen also causes the neurons in this region to express receptors for progesterone (estrogen binds to the promoters of the progesterone receptor gene to turn on their transcription). Then, when progesterone surges, slightly later, it bonds progesterone receptors, and this causes the female to seek out males, present her genitals, and engage in other come-hither behaviors (such as ear wiggling in rats). The female's ventromedial nucleus is integrating two types of information. One is electrical, triggered by sensory stimulation from seeing/hearing/smelling the male (hey, he's kinda cute . . .). The other is hormonal information indicating her ovarian status (. . . and I'm fertile right now!). Only when both signals are on will things move ahead. Recordings of neuronal activity from the ventromedial nucleus show that neurons are firing spikes at high rates both during this "courtship phase" and during subsequent copulation. Females whose ventromedial nucleus has been damaged will not show these behaviors even if their ovarian hormones are functioning normally. Conversely, artificial electrical stimulation of this nucleus can induce or strengthen female-typical mating behavior.

Estrogen also functions to trigger the cells lining the vagina to produce substances that ultimately result in an odor that is attractive to males. These odorant molecules do not appear to be directly secreted by the cells of the vagina, but rather by bacteria that thrive in an estrogen-primed environment within the vaginal mucus. These odors are key to triggering sexual interest in males. The vaginal odor of a monkey in the postovulatory phase not only is unattractive to males; it even seems to repel them. If vaginal secretions from a female monkey in heat (immediate preovulatory phase) are smeared around the vagina of a monkey not in heat, the male will be fooled by the alluring smell and will attempt to mate.

Here, it is worth pausing to mention that although some general themes of this sexual circuit are likely to operate in human females, there are important differences as well. As noted previously, olfaction does not appear to be as central to sexual behavior in humans as it is for many other mammals. Likewise, ovarian hormones do not exert such a powerful control over female sexual drive in women. In fact, women who have had their ovaries removed for medical reasons typically experience a normal sex drive.

Males also have a center in the hypothalamus for triggering sexual behavior, but this is in a different area, the medial preoptic region. This is a group of nuclei within the hypothalamus that includes the aforementioned INAH3, the area that has a high density of testosterone receptors and that is larger in males. Like the ventromedial nucleus in females, the medial preoptic region integrates both sensory-driven synaptic stimulation from higher centers, including emotional centers and hormonal information. The difference is that in this case the hormone is testosterone. If the action of testosterone is removed (by castration or drugs that block testosterone receptors), then this will block the increase of spike firing of medial preoptic area neurons evoked by sexual stimuli, such as a female in heat. These treatments will also cause a reduction in male-typical sexual behavior, such as the mounting of females. A complete abolition of the mounting behavior can come from selective destruction of the medial preoptic area. Surprisingly, this does not seem to produce a complete abolition of sex drive, but only stops that triggered by females: male monkeys with medial preoptic lesions will still masturbate with gusto.

Artificial electrical stimulation of a male monkey's medial preoptic area will cause it to mount a nearby female and commence thrusting, but copulation will only be sustained if the female is in heat. Otherwise, the male will give a few half-hearted thrusts and scamper off. It is important to realize that the medial preoptic area, which is quite small, triggers penile erection, mounting, and thrusting, but it is not the actual command center for any of those actions. Rather, it activates brainstem centers to produce erection and motor cortex and motor coordination centers to initiate mounting and thrusting.

Likewise, the medial preoptic area does not appear to be important in triggering ejaculation. Artificial stimulation of this region will not result in ejaculation, and recordings of electrical activity do not show a burst of activity correlated with ejaculation, as would be expected if the medial preoptic were the center for triggering this function. In fact, the medial preoptic falls almost completely silent at the point of ejaculation and remains so for minutes afterward. It has been suggested that this may underlie the male post-ejaculation refractory period during which further sexual activity is difficult or impossible.

This brings us to the topic of orgasm. Orgasm, as a physiological phenomenon, is remarkably similar in women and men. In both sexes, orgasm involves a rising heart rate, an increase in blood pressure, involuntary muscle contractions, and an intensely pleasurable sensation. Orgasm is accompanied by contraction of two pelvic muscles, the bulbocavernosus and aschiocavernosus, as well as the muscles in the wall of the urethra, leading to ejaculation of semen in men, and in some cases, glandular fluids in women as well (one recent survey indicated that 40 percent of women have experienced ejaculation at some point).

In recent years, brain imaging studies have been performed on men during orgasm. Let's think for a moment about what a profoundly unsexy arrangement that is. The subject has his head immobilized with a tightly fitting strap that is then slid into the ringing, claustrophobic metal tube that is the positron emission tomography (PET) scanner, with his nether regions still outside the device. An intravenous line is attached to deliver a pulse of the radioactive water required for PET imaging. The subject is instructed to close his eyes and lie as still as possible (to avoid activation of visual or motor parts of the brain) while his female companion attempts to bring him to orgasm with manual stimulation. It's amazing that anyone could achieve orgasm under these conditions. Yet, in a recent study by Gert Holstege and his coworkers at University Hospital Groningen in The Netherlands, 8 of 11 subjects were able to ejaculate during the experiment (and 3 of these even managed to do it twice).

During male orgasm, a large number of brain regions were activated. Predictably, the reward centers of the midbrain, including the ventral tegmental area, were strongly engaged. In this sense, both new love and orgasm are like heroin and cocaine in their pleasurable effects. A large number of discrete areas in the cortex were activated including sites in the frontal, parietal, and temporal lobes. Surprisingly, these sites of cortical activation were found only on the right side of the brain. Finally, the cerebellum was also strongly activated during orgasm. This is not entirely unexpected because part of the job of the cerebellum is to detect mismatches between plans for motor action and feedback about how that is progressing. In this sense, involuntary motions produced during orgasm might be expected to produce strong cerebellar activation. Although imaging studies of female orgasm have not yet been been published at the time of this writing, preliminary results presented at scientific meetings have indicated that the patterns of brain activation in female and male orgasm are remarkably similar. The main difference between women's and men's brains during orgasm appears to be that women have additional strong activation in a midbrain area called the periaqueductal gray region. This is an area rich in endorphin-containing neurons and might possibly contribute an additional aspect to the sexual pleasure or satiety felt by women.

Orgasm is a complex phenomenon with different aspects mediated by different brain regions. This is revealed in part by brain stimulation studies in which electrical activation of the septum (a part of the limbic system subserving, among other things, emotion and memory) produced orgasms in men that did not have any pleasurable component to them. Similarly, there are a number of cases of patients suffering from seizures involving the right temporal lobes (in which parts of the limbic system reside) who experience uncontrollable seizure-evoked orgasms with no pleasure. But it's important to note that not all seizure-evoked orgasms fail to evoke sexual pleasure: Yao-Chung Chuang and his coworkers from Chang Gung Memorial Hospital in Taiwan report the case of a 41-year-old woman who experienced temporal lobe seizures and accompanying pleasurable orgasms after a few seconds of toothbrushing (Figure 6.4). Gives the expression "oral sex" a whole new meaning, doesn't it? It's tempting to speculate that this woman's seizures activated the midbrain reward circuitry, including the ventral tegmental area, while those who experienced seizure-evoked orgasms without pleasure did not undergo activation of that circuitry.

The observation that it is possible to have orgasms without pleasure is remi-

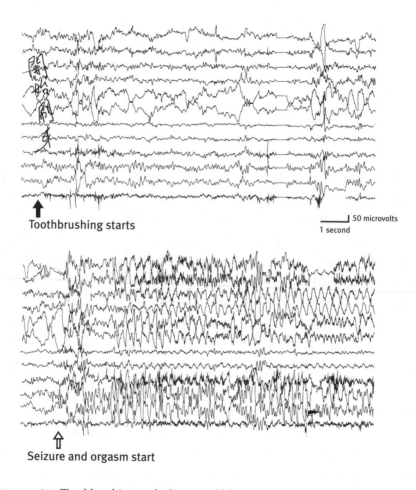

Toothbrushing starts

50 microvolts

1 second

Seizure and orgasm start

FIGURE 6.4. Toothbrushing-evoked temporal lobe seizures in a 41-year-old woman from Taiwan. These seizures, shown here in an electroencephalogram (EEG) record, were accompanied by pleasurable orgasms. Reproduced with the permission of Elsevier from Y.-C. Chuang, T.-K. Lin, C.-C. Lui, S.-D. Chen and C.-S. Chang, Tooth-brushing epilepsy with ictal orgasms, *Seizure* 13:179–182 (2004).

niscent of a theme we have previously encountered in sensory systems: separate brain regions are involved in the pure sensory aspects of an experience and the emotional (rewarding/aversive) components. Normally, these strands of sensation are tightly woven together and it is only when something unusual happens, such as the onset of Capgras syndrome or pain asymbolia (Chapter 4) or orgasmic seizures without pleasure, that we can see the underlying component parts.

Aside from the intense and immediate pleasure of orgasm, there is also the warm, lingering post-orgasmic afterglow. This state, which is thought to be crucial for sexual pair bond formation, may be mediated, in both men and women, by release of the hormone oxytocin from the pituitary gland, under control of the hypothalamus. Treatments that block oxytocin release do not prevent orgasm or the immediate pleasurable sensations, but do seem to interfere with the crucial afterglow. It is worth noting that the oxytocin-releasing system appears to be involved in more general aspects of pair bond formation, not just that which occurs in a sexual context. Oxytocin surges occur in mothers' brains at birth and during breastfeeding and are likely to be one important factor in developing a mother's bond with her child.

WE TEND TO use a shorthand to describe the sexual feelings and motivations of individuals. Typically we will say that someone is homosexual, heterosexual, or bisexual as a descriptor of what we have come to call sexual orientation. In truth, this is a very crude sort of measure. In humans, sexuality has been embellished considerably beyond instinctive behavior. Each of us carries within his or her mind a sort of template of an ideal romantic or sexual encounter that incorporates many unique elements and details. Within each of these categories there is a lot of variation. In the gay community, for example, one can find gay men and lesbians with all sorts of gender identities. Lesbians run the gamut from "butch" to "femme," with many who have sexual and gender-identity feelings that are not easily classifiable along this dimension. The playwright and actor Harvey Firestein has described himself as "gay as a pink leather piñata." This is a hoot, but what does it mean? Macho-gay? Effeminate-gay? Neither? Straight people are similarly diverse, adopting myriad sexual roles and personas.

The subtlety of human sexual identity makes it hard to analyze. But within these rough designations it seems that, at least in the United States and Europe, about 4 percent of men and 2 percent of women are consistently homosexual, about 1 percent of men and 2 percent of women are consistently bisexual, with the remainder heterosexual. These numbers are based on surveys and it is not always easy to get truthful answers without sampling bias, but these are reasonable general estimates that have been confirmed in multiple carefully controlled studies. They reflect consistent behavior, not "experimentation"—the numbers of people who have ever had at least one homosexual experience leading to orgasm is much higher (approximately 25 percent for men and 15 percent for women).

The biological determinants of sexual orientation have been a topic of rancorous, politically charged debate that has intensified in recent years as more and more science bearing on this issue has come to light. Many religious conservatives and some others on the political right are invested in the interpretation that homosexuality is a sinful choice made of free will. As a consequence they have been motivated to attack any research suggesting that sexual orientation has a biological component whether genetic or epigenetic, driven by nongenetically determined biological signals such as fetal hormone levels. Gay activists and many on the political left wish to promote societal acceptance of and civil rights for gays. Accordingly, many in this camp would like to believe that sexual orientation is like eye color: it's a trait you're born with rather than a choice. There is also a flip side of fear to this position, however. If sexual orientation is completely genetically determined, then one must be concerned that in the future people might use genetic tests to discriminate against gays or even abort potentially gay fetuses.

Let's try to be as objective as possible in examining the evidence to date. As in any nature-versus-nurture type of debate, one can take extreme positions, but these are not inevitable. Recall the discussion (in Chapter 3) of the biological basis of general intelligence, another contentious and complex human trait that's hard to measure. In that case it appears likely that about 50 percent of general intelligence is heritable. It's possible that, in the fullness of time, a similar answer will emerge for sexual orientation.

So, does sexual orientation have a heritable component? Statistically, having a gay sibling dramatically increases your own probability of being gay. It appears that about 15 percent of the sisters of lesbians are themselves lesbian (compared to about 2 percent of the general population), and 25 percent of all brothers of gay men are gay (compared to about 4 percent of the general population). Interestingly, having a gay brother does not increase the chance of a woman's being a lesbian and vice versa. Now, of course, the studies that produced these results do not speak directly to the heritability of sexual orientation since siblings also share similar upbringing and environment. More compelling evidence comes from studies of monozygotic (identical) and dizygotic (fraternal) twins. Here, it seems that, for men, having a gay male monozygotic twin makes your own likelihood of being gay about 50 percent while having a gay male dizygotic twin makes your likelihood of being gay about 30 percent (similar to the likelihood when you have a gay nontwin brother). A similar study conducted with women showed that having a lesbian monozygotic twin conferred a 48 percent chance of being lesbian, while a lesbian dizygotic twin was associated with a 16 percent rate (again, similar to that from having a nontwin lesbian sister).

There is one clear conclusion from these studies: In significant numbers of cases, monozygotic twin pairs are discordant (one gay, one straight), which indicates that sexual orientation is not 100 percent heritable, like eye color. That said, the studies suggest that a portion of sexual orientation *is* genetically determined. But we have to be concerned about the limitations on studies of twins raised together: if monozygotic twins are raised more similarly than dizygotic twins, this could contribute to the greater incidence of homosexuality in the former. A better study, of course, would analyze twins raised apart. These are ongoing at the time of this writing.

Two lines of evidence suggest that male homosexuality is partially linked to function of the X sex chromosome, which males inherit from their mothers. A small number of men have an extra X chromosome, giving them the XXY genotype instead of the more typical XY. This is called Klinefelter's syndrome and has a number of associated traits, including reduced levels of testosterone and reduced sperm viability. In one study, such men had a much higher incidence of homosexuality than the general population (about 60 percent). A complementary line of work has shown that among chromosomally normal gay men, significantly increased rates of same-sex orientation were found in the maternal uncles and male cousins of these subjects, but not in their fathers or paternal relatives. This is also consistent with maternal transmission through the X chromosome.

Taken together, these studies indicate a strong but not total genetic influence on sexual orientation in both men and women, with the effect in women being somewhat smaller. What gene or genes are likely to be involved? Here it is worthwhile to briefly discuss some issues of genetics in relation to human behavior. Complex human behavioral traits such as general intelligence, shyness, and sexual orientation can have a significant degree of heritability, yet these traits cannot typically be attributed to variation in a single gene. Rather, they are polygenic: the heritable component of the trait is caused by variation at multiple genes. As a fictional illustrative example, we might imagine that general intelligence is promoted by having a cerebral cortex composed of a large number of highly interconnected neurons that fire spikes readily. Thus higher general intelligence would be promoted by particular forms of genes that increase the overall number of neurons in the cerebral cortex, others that promote the growth and branching of dendrites or axons, and yet others that express ion channels that underlie particular modes of spike firing. Given the broad activation of brain regions involved in amorous and sexual behavior that we have seen in imaging studies, we can imagine that sexual orientation is also likely to have polygenic heritable components.

The influence of maternal inheritance on male homosexuality makes the X chromosome a reasonable place to look for one or more genes that could influence male sexual orientation. Dean Hamer and his colleagues at the National Institutes of Health examined the DNA of a group of gay men and lesbians who had at least one same-sex homosexual sibling, as well as a control group of straight men and women. They did this by analyzing stretches of DNA at roughly evenly spaced locations throughout the X chromosome. They found that a particular region of this chromosome, called Xq28, had a significant tendency to differ from that of straight men in gay men, but not in lesbians. This finding does not pinpoint a particular gene. Rather, it points to the likelihood that one or more genes in this chromosomal region may vary in a way that contributes to male homosexuality. More recently, this type of genetic scan was performed on gay male DNA but instead of just looking at the X chromosome, the experimenters analyzed markers spread across the whole genome (all 23 chromosomes). Several additional "linkage sites" were found on chromosomes 7, 8, and 10. It is important to note that as of this writing, no group has published a scientific paper replicating these linkage studies of the Hamer lab, an event that would go a long way toward validating this notion of particular genetic loci influencing homosexuality.

Genetic variation may not account for all of the biological component of sexual orientation. It's possible that epigenetic developmental factors may also contribute. These might include the effects of maternal stress or immune system status during pregnancy and hormonal effects derived from siblings in utero. The latter seems to be an important factor in rats: female pups that are adjacent to male siblings in the womb sometimes are partly masculinized, both physically and behaviorally, as a result of testosterone circulating from their brothers in utero. Although this may also be an issue in human multiple pregnancies, it is likely to be less dramatic because the maternal blood supply to the fetuses is more separated in humans, whereas in rats it is organized serially such that a particular pup is "downstream" from another.

How might genetic or epigenetic developmental factors work to influence sexual orientation? The basic hypothesis has been this: Gay men have brains the structure and function of which are, in some respects, like those of straight women. Conversely, lesbians have brains that are, in some respects, like those of straight men. So, one obvious test of this hypothesis is to look at those regions that were previously known to be structurally different in the brains of straight men and women. This is exactly what Simon LeVay of the Salk Institute did. He measured the volume of the hypothalamic nucleus INAH3 in postmortem tissue samples of straight and gay men as well as of straight women. The gay sample was entirely composed of men who had died of AIDS, and the straight sample was partly men who had died of AIDS (intravenous drug users) and partly men who had died of other causes, as well as women who had died of other causes. Replicating previous work from another lab, LeVay found that the volume of INAH3 was two to three times larger in the straight men than in the straight women. The really interesting finding was that the average volume of INAH3 in the gay men was similar to that of straight women: two to three times smaller than that of straight men. These differences were not seen in adjacent hypothalamic nuclei that are not sexually dimorphic in straight people, such as INAH 1, 2, and 4.

Is it possible that the INAH3 was smaller in the gay male sample because of AIDS, which is known to affect brain cells? This is unlikely because the mean INAH3 volume of the AIDS-infected straight male sample was also significantly larger than the INAH3 volume of the gay sample. In addition, after his initial publication in 1991, LeVay was able to get brains from some gay men who had died of causes other than AIDS, and he found the same differences as in the previous group.

Another study looked at the anterior commissure, which, you will recall, is larger in straight women than in straight men. The cross-sectional area of this bundle of axons connecting the right and left sides of the brain was measured in postmortem brains from gay men, straight men, and straight women by Laura Allen and Roger Gorski at UCLA. They found that gay men had anterior commissures that were larger, on the average, than those of straight men and even slightly larger than those of straight women.

These anatomical results with INAH3 and the anterior commissure have received a lot of attention, much of it overblown. Newspapers and magazines around the world have rushed to declare that these data prove that "homosexuality is genetic" or that "gay people are born that way." Clearly, a correlational study of adults cannot prove this sort of statement. Although these studies are consistent with the notion that sexual orientation is, at least in part, biologically determined, we still don't know the answer to the crucial question: What are the brains of gay people like at birth or shortly thereafter, before socio-cultural factors have a chance to make a major impact?

If gay people are indeed "born that way," then one might expect that the masculinization of the female brain and the feminization of the male brain will be evident in nonsexual behavior and physiology, starting early in life. One approach for assessing this idea is to interview people and their relatives and friends about their recollections of childhood to see if particular themes emerge in the homosexual population. This strategy is fraught with difficulty, however, because it relies on people's memories and it is very difficult to exclude sampling bias. Nonetheless, it is interesting that gay men as a population tend to have strong recollections of effeminate behavior in early childhood. In fact, a study by James Weinrich and his coworkers at the University of California at San Diego showed that the strongest recollections of effeminate childhood behavior in gay adults were found in that part of the gay male population that adopted the most female-typical role in their adult lives (for example, strongly preferring the receptive role in anal intercourse). This amplifies the point made earlier: Straight, gay, and bisexual are crude categories and their use in genetic, anatomical, and behavioral studies may mask some of the most interesting findings. One could imagine, for example, that "macho" gay men have a larger INAH3 and smaller anterior commissure than effeminate gay men or that "femme" lesbians have a smaller INAH3 and larger anterior commissure than "butch" lesbians. These subcategories of homosexual behavior are themselves quite crude, but I think they make the point: graded differences in certain brain structures may be manifest in part as subtle differences in sexual orientation.

A much more powerful method than the retrospective interview is the prospective study, like that carried out by Richard Green of Imperial College School of Medicine, who identified boys with effeminate behaviors in the preschool years. When these boys were then tracked as they grew up, it was found that greater than 60 percent of them became gay or bisexual adults. This is a remarkable statistic when you consider that homosexual plus bisexual men constitute only about 5 percent of the adult male population.

Another prediction of a biological hypothesis for sexual orientation is that manipulations that feminize male brains or masculinize female brains increase the incidence of homosexual behavior. This seems to be the case in both human and animal studies. Male rats with lesions in the medial preoptic area (including INAH3) often engage in female-typical sexual behaviors toward other males, such as ear wiggling and lordosis, a posture that presents the genitals. This effect can be further enhanced if the lesioned males are treated with estrogen. Similar effects can be produced by treatments that impede the action of testosterone (castration at birth, drugs that interfere with testosterone receptors). Most interestingly, subjecting a mother rat during pregnancy to moderate stress (confinement in a clear plastic tube under bright lights) can reduce the levels of testosterone in the developing fetus. When the male pups grow up, their sexual behavior is feminized: they are reluctant to mount females and themselves display female-typical sexual behavior. In other words, interfering with testosterone in fetal or early postnatal life can make male rats "gay."

Not surprisingly, a complementary result is found in females exposed to higher than usual levels of testosterone. The female offspring of rats or sheep given testosterone-boosting treatments in utero tend to adopt more male-typical sexual behaviors (mounting, aggression). More important, women who suffer from congenital adrenal hyperplasia, in which testosterone levels are elevated starting in utero, have a much higher incidence of lesbianism than is found in the general population.

One of the bitterest issues in the ongoing debate about the biological basis of sexual orientation has been whether gay men and lesbians can change their sexual feelings and behavior to become straight. Some researchers, such as Robert Spitzer of the New York State Psychiatric Institute, have published papers claiming that with certain forms of treatment this is possible in at least a fraction of the population (17 percent of the males in his sample reported achieving "exclusively opposite sex attraction" after treatment). Others, including the major professional associations of clinical psychologists and psychiatrists, have derided these claims as politically motivated junk science, in part based on a critique of Spitzer's sampling methods, in which subjects were recruited from ex-gay ministries. Although the advisability of offering treatment to change sexual orientation is itself an important moral and social question, whether or not some people can change their sexual behavior from gay to straight does not bear on the question of whether sexual orientation is, to some degree, biologically determined. Left-handedness is almost certainly biologically determined and yet almost any leftie can be trained to be right-handed. Some Catholic priests and others, who have normal sexual drives, can subsume these feelings to completely refrain from sexual acts, in keeping with religious teachings. So, even if a small fraction of homosexuals can undertake a form of treatment that results in their adopting exclusively heterosexual behavior, this does not speak to the question of whether sexual orientation is, either fully or in part, determined by biological factors at birth or shortly thereafter.

AT THIS POINT, the evidence from families, twin studies, gene linkage, neuro-anatomical analysis of postmortem tissue, and manipulations of sex hormones points to the conclusion that some portion of sexual orientation is biologically determined. Whether this will turn out to be 30 percent or 90 prcent of the variation in sexual orientation remains to be seen. Likewise, the relative contribution of genetic and epigenetic factors to this biological predisposition remains unclear. It is likely that, many years from now, when the smoke clears, sexual orientation will have a story not very different than that emerging for many other complex human behaviors: It will have some degree of sociocultural and some degree of biological determination. The biological part will have both genetic and epigenetic components and the genetic component will reflect the action of multiple genes.

Chapter Seven

Sleeping and Dreaming

IT WAS 1952 AND the American military was getting panicky. Over 60 percent of the airmen captured by the Chinese army in the ongoing Korean War
were confessing to bogus war crimes (such as the use of biological weapons) or
were signing statements or recording messages renouncing the United States
and embracing communism. These events created an enormous propaganda
coup for the Chinese. The CIA and military intelligence specialists entertained
a number of theories about the success of this effort, including the development of exotic "brainwashing drugs," hypnosis, and exposure to mind-altering
electric fields. The truth, revealed some years later, was much more prosaic:
the Chinese were able to coerce these statements from their prisoners mainly
through the use of beatings combined with prolonged sleep deprivation.

This shouldn't have been news. Throughout history it has been known that sleep deprivation is an ideal form of torture. The ancient Romans employed sleep deprivation extensively to interrogate and punish prisoners. It leaves no physical trace and it does not result in permanent alteration of the victim's mental function: he or she is mostly back to normal after a good night's sleep or two. Indeed, of the thousands of American and United Nations prisoners of war in Korea, almost none maintained their bogus confessions or denunciations after their release. They weren't "brainwashed" at all. Their fundamental belief systems and personality had not been permanently compromised. Rather, they were temporarily rendered delusional, suggestible, and even psychotic through sleep deprivation.

In his autobiographical book *White Nights,* the Russian dissident Menachem Begin, later to become the prime minister of Israel, describes his sleep-deprivation treatment at the hands of the KGB.

> In the head of the interrogated prisoner, a haze begins to form. His spirit is wearied to death, his legs are unsteady, and he has one sole desire: to sleep . . . Anyone who has experienced this desire knows that not even hunger and thirst are comparable with it.
>
> I came across prisoners who signed what they were ordered to sign, only to get what the interrogator promised them.
>
> He did not promise them their liberty; he did not promise them food to sate themselves. He promised them—if they signed—uninterrupted sleep! And, having signed, there was nothing in the world that could move them to risk again such nights and such days.

Begin's description highlights the effectiveness, but also the limitations, of sleep deprivation as a torture method. It is very effective as a coercive device, but torturers cannot rely upon information gleaned while a prisoner is in a severely sleep-deprived state: people in this condition are often experiencing auditory and visual hallucinations as well as paranoia. They are likely to say anything if they believe they finally will be allowed to sleep. I should note that torture by sleep deprivation is a practice that is still in common use. Andrew Hogg of the Medical Foundation for the Care of Victims of Torture in the United Kingdom says, "It is such a standard form of torture that basically everybody has used it at one time or another." Here, "everybody" includes democratic states such as the United States, the United Kingdom, India, and Israel, all of which have published recent guidelines for interrogation by the military and security services that allow for extreme sleep deprivation.

How long can a human go without sleep? The world record is presently held by Randy Gardner, who, as a 17-year-old high school student, stayed awake for 11 straight days in 1965 just for the hell of it. He did this without the use of stimulant drugs. During this period, Gardner initially became moody, clumsy, and irritable. As time progressed he showed delusions (he said that he was a famous professional football player), then visual hallucinations (he saw a path through a forest extending from his bedroom), paranoia, and a complete lack of mental focus. Remarkably, after a 15-hour sleep, almost all of these symptoms abated. Gardner appears to have suffered no lasting physical, cognitive, or emotional harm from the incident.

A grisly set of experiments with rats showed that total sleep deprivation will cause death in 3–4 weeks. Although the exact cause of death was unknown, these animals suffered from skin lesions and a gradually failing immune system. This condition ultimately allowed for the colonization of the body by otherwise benign bacteria that are usually restricted to the digestive tract. Throughout this period there is a gradual buildup of the steroid hormone cortisol, a natural immunosuppressant, and a gradual reduction in core body temperature. Human death from total sleep deprivation has not been reported in the scientific literature. But there are indications of this from records of Nazi death camp experiments during World War II as well as reports of executions by sleep deprivation in China in the nineteenth century. These suggest that 3–4 weeks of sleep deprivation will kill humans as well. Or, to put it another way, 4 weeks without food may or may not kill you (depending upon your health, age, and access to medical care) but 4 weeks without sleep will.

CLEARLY, BOTH RATS and humans need sleep to live. This raises the question: what are the physiological functions of sleep that make it so important? Amazingly, we don't have a definitive answer to this simple question. One obvious idea is that sleep serves a restorative function for the entire body. Cellular growth and repair functions involving gene expression and protein synthesis seem to accelerate during sleep in both the brain and other tissues. But it is not well established that people who are physically active sleep significantly more than those who are confined to bed. Nor is it clear that a brief period of intense exercise promotes longer total sleep time (although there are some small effects on the time spent in various stages of sleep).

It has been proposed that sleep functions to conserve energy. This may be particularly relevant for warm-blooded animals (mammals and birds) that must expend a lot of energy to maintain a body temperature higher than that of their surroundings. Indeed, many small mammals living in cold climates, who lose heat easily by having an unfavorable surface area to body weight ratio, tend to sleep a lot, often in insulating burrows. Yet sleep does not appear to have evolved only in warm-blooded animals. EEG recordings from reptiles and amphibians indicate that they also sleep, and there are now strong indications of a sleep-like state in some invertebrates, such as crayfish, fruit flies, and honey bees. Also, though it is true that the overall use of energy is reduced during sleep, as compared with the active waking state, there is almost as much reduction in energy use from just resting quietly. The additional energy conservation in going from the resting state to sleep is minimal. So, an explanation for sleep based on restoration and energy conservation is unlikely to be complete.

One simple role of sleep might be to restrict an animal's activity to those times when activity is productive—when the chance of finding food is high but the chance of becoming someone else's food is low. For many species, including ours, this means sleeping at night. Others, such as many foraging rodents, bats, and owls, do the opposite, but the principle in the same: they are trying to hunt for food but avoid predators. There is some evidence to support this model: mammals at the top of the food chain such as lions and jaguars tend to sleep a lot (as much as 12 hours a day) while those that graze in the open such as deer and antelope sleep much less. Some herbivorous animals such as ground squirrels and sloths also sleep a lot (two-toed sloths sleep for 20 hours every day!), but these tend to be species that are mostly safe from predation during sleep because of their sleeping location (in underground burrows or high in trees). Nonetheless, this explanation for sleep doesn't seem entirely satisfying either. Perhaps if we look at the process of sleep in greater detail, more compelling ideas will emerge.

THE SCIENTIFIC STUDY of human sleep has a very strange beginning. In the nineteenth century, several investigators in France were very interested in the processes of sleep, but they never did the most simple, observational experiment: just stay up all night and make notes of how people's bodies move over the course of a normal night's sleep. Instead, these scientists spent their time trying to influence the dreams of their subjects. They would open a bottle of perfume under the sleeper's nose or tickle him with a feather and then wake him up a few minutes later to see if they had influenced his dreams. Not much useful information came from this line of work, and up until the 1950s the standard model of sleep was simple and wrong. It was held that sleep is a constant, unchanging period of little body movement and low brain activity that changes only upon waking.

In 1952 Eugene Aserinsky was a graduate student in the laboratory of Nathaniel Kleitman at the University of Chicago, where EEG recordings were being made from adults as they fell asleep. These revealed that after falling asleep, the EEG gradually changed from a desynchronized, low-voltage trace to a high-voltage trace with slow, synchronized oscillations. At this point, it was assumed that deep sleep had been achieved and this status would be maintained until waking. The standard operating procedure was to record for 30–45 minutes to capture this transition and then turn the EEG recorder off to save chart paper. One night Aserinsky brought his son Armond, 8 years old, into the lab to be the subject. About 45 minutes after Armond had fallen asleep, his father was watching the pens on the EEG chart recorder register the slow oscillations of deep sleep. Then, amazingly, the EEG shifted to another rhythm that looked more like waking even though Armond was still clearly sleeping and was totally immobile. We now know that this stage of sleep is associated with rapid eye movements (REMs) and that while it usually does not occur in adults until about 90 minutes after falling asleep, in children, like Armond, it occurs sooner.

The report of these findings by Aserinksy and Kleitman in 1953 began the modern era of sleep research, and in the following years a much more detailed picture of sleep emerged. When scientists left their EEG machines on all night (piling up enormous stacks of chart paper in the process), they found an adult sleep cycle of about 90 minutes duration (Figure 7.1). This consisted of the aforementioned gradual descent into deeper and deeper sleep accompanied by gradual synchronization of the EEG. These stages of sleep are collectively called non-REM sleep and they are further subdivided into four stages ranging from

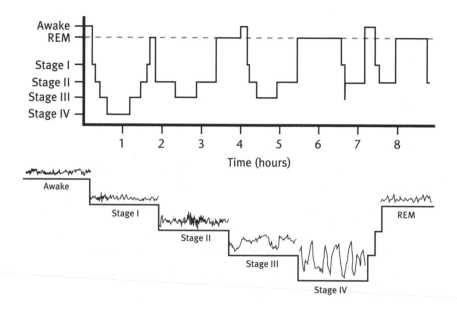

FIGURE 7.1. The stages of adult human sleep. The top panel depicts a complete night's sleep with sleep stage on the vertical axis. This graph was made by analyzing the EEG record to determine the sleep stage. It shows the main features of a normal night's sleep. There is a sleep cycle of approximately 90 minutes duration during which the sleeper gradually progresses from drowsiness (stage I) into deep sleep (stage IV), followed by a period of REM sleep. A typical night's sleep might involve 4 or 5 of these cycles. As the night progresses, a higher proportion of the sleep cycle is devoted to REM sleep with a concomitant decrease in non-REM sleep (stages I–IV). The bottom panel shows representative EEG records from each sleep stage. Note that the EEG record for REM sleep is similar to that of the waking or drowsy state. Adapted with the permission of Macmillan Publishers, Ltd., from E. F. Pace-Schott and J. A. Hobson, The neurobiology of sleep: genetics, cellular physiology, and subcortical networks, *Nature Reviews Neuroscience* 3:591–605 (2002). *Joan M. K. Tycko, illustrator.*

drowsy/nodding-off (stage I) to deep sleep (stage IV). A typical uninterrupted night's sleep will consist of 4 or 5 complete 90-minute-long cycles. What's interesting is that as the night wears on, the character of each sleep cycle changes so that there is proportionally more REM and less non-REM sleep per cycle. In the last period before waking, as much as 50 percent of the cycle may be devoted to REM sleep.

It's a testament to the occasional bone-headedness of scientists that sleep cycles were not discovered before the 1950s. You don't need an EEG recording to detect them. Simple observation of a sleeper throughout the night will show you the main features. The most obvious of these is the rapid side-to-side eye movements that are easily seen even when the eyelids are closed (owing to the bulge of the cornea indenting the eyelid). Careful observation would reveal a host of other changes during REM sleep. These include an increase in breathing rate (as well as heart rate and blood pressure) and a sexual response (penile erection in men, erection of the nipples and clitoris together with vaginal lubrication in women). Even more striking are changes in muscle tone. The typical adult sleeper will change his or her position about 40 times a night without being conscious of this action. None of these motions, however, will occur during REM sleep. In REM sleep, there is no movement at all. In fact, there is not even any muscle tone: the body goes totally limp. It is almost impossible to have REM sleep in anything other than a horizontal position. Remember this the next time you are wrapped in an airline blanket and stuffed into your coach class seat like a scrofulous burrito for a trans-Atlantic flight: even if you manage to catch some sleep in your seat, you won't be able to enter REM sleep.

REM sleep is sometimes called "paradoxical sleep" because the EEG resembles that of the waking state, yet the subject is essentially paralyzed. The story here is that the motor centers of the brain are actively sending signals to the muscles but these signals are blocked at the level of the brainstem by inhibitory synaptic drive. This blockade affects only the outflow of motor commands down the spinal cord, not those of the cranial nerves that exit the brainstem directly to control eye and facial movements (as well as heart rate). Michel Jouvet of the University of Lyon showed that severing the inhibitory fibers that block motor outflow in cats resulted in a bizarre condition: during REM sleep the cats engaged in complex motor behaviors while keeping their eyes closed. They ran, pounced, and even seemed to eat their imagined prey. Although we can't know this for certain, they appeared to be acting out their dreams (more on this soon). A similar phenomenon is seen in a human condition called REM sleep behavior disorder, which mostly affects men over 50. This disease causes dream-enacting behaviors during the REM period of sleep, including kicking, punching, jumping, or even running. Not surprisingly, these violent behaviors can often result in injury to the patient or to his or her bedmate. In most cases, this disorder is successfully treated by a bedtime dose of the drug clonezepam (sold under the trade name Klonopin), which works by boosting the strength of synapses that use the inhibitory neurotransmitter GABA. REM sleep behavior disorder is different from conventional sleepwalking, which occurs only during non-REM sleep.

Humans show changes in sleep over the life cycle, with the proportion of the time spent in REM sleep decreasing from about 50 percent at birth to 25 percent in mid-life and 15 percent among the elderly (a decrease in REM is also seen over the lifespans of cats, dogs, and rats). If we compare our sleep with that of other mammals, we find that we are more or less in the center of the range bounded by the duck-billed platypus, which spends about 60 percent of its sleeping life in REM, and the bottlenose dolphin, which has a REM proportion of only 2 percent. There is no obvious relationship between degree of REM sleep and brain size or structure across mammalian species (Figure 7.2). Non-REM sleep appears to have evolved as early as the fly (about 500 million

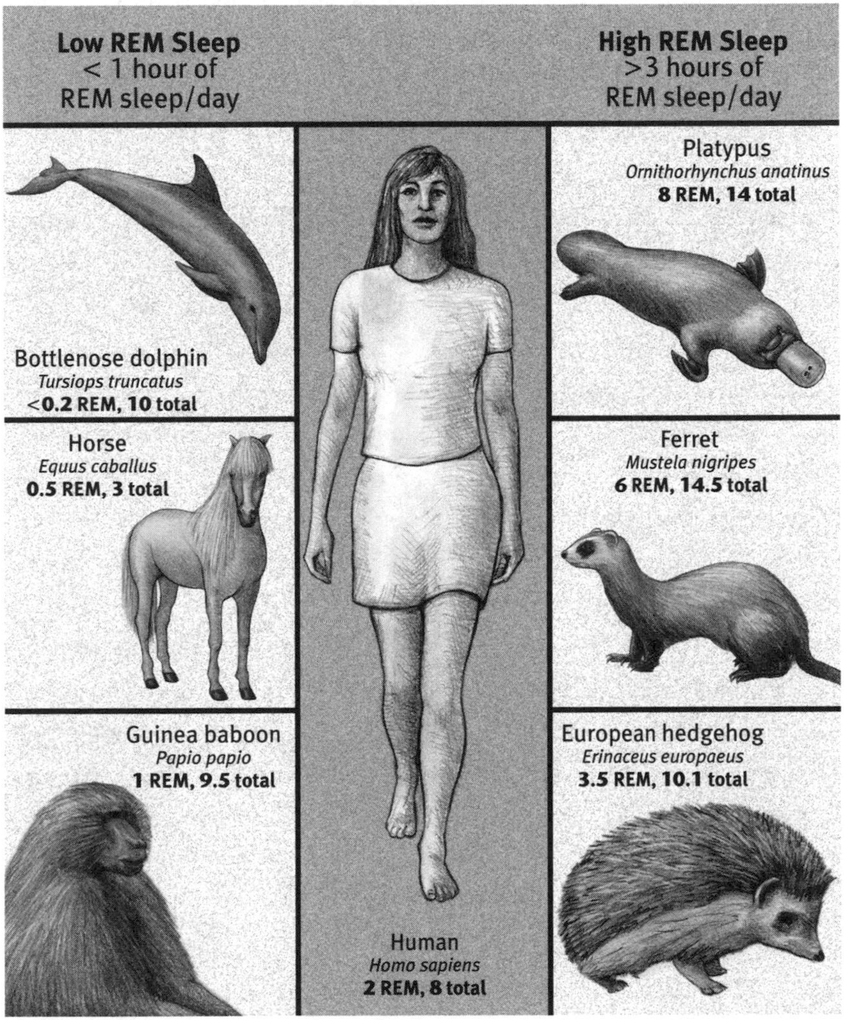

Low REM Sleep
< 1 hour of
REM sleep/day

High REM Sleep
>3 hours of
REM sleep/day

Bottlenose dolphin
Tursiops truncatus
<0.2 REM, 10 total

Horse
Equus caballus
0.5 REM, 3 total

Guinea baboon
Papio papio
1 REM, 9.5 total

Platypus
Ornithorhynchus anatinus
8 REM, 14 total

Ferret
Mustela nigripes
6 REM, 14.5 total

European hedgehog
Erinaceus europaeus
3.5 REM, 10.1 total

Human
Homo sapiens
2 REM, 8 total

FIGURE 7.2. REM and total sleep in a gallery of representative mammals. Humans fall into the middle of the range when REM sleep is considered either as a raw value or as a proportion of total sleep. Adapted from J. M. Siegel, The REM sleep-memory consolidation hypothesis, *Science* 294:1058 1063 (2001); copyright 2001 AAAS. *Joan M. K. Tycko, illustrator.*

years ago), but true REM sleep is found only in warm-blooded species. It is present in the most primitive surviving mammals (such as the platypus and the echidna) as well as in birds, but appears to be absent in reptiles and amphibians.

So, with knowledge of sleep cycles, we can return to our main question "Why is sleep necessary?" with a bit more sophistication. Really, two separate questions are warranted: What are the key functions of sleep composed of only a non-REM period, as is found in reptiles and amphibians (and possibly some invertebrates as well)? And, what are the key functions of cycling sleep in which REM and non-REM periods alternate, as is found in mammals and birds? It may be that the previously mentioned ideas that sleep is required for restorative functions, energy conservation, and maximizing feeding efficiency while minimizing danger from predation are appropriate for non-REM sleep alone. Cycling sleep is serving some function that only emerges in mammals and birds and that is most important early in life. Let's consider some hypotheses about what that function might be. One proposal is that cycling sleep serves a rather mundane function. It's known that non-REM sleep tends to cool the brain, reducing its thermoregulatory set point, while REM sleep heats the brain up. Perhaps alternating bouts of REM and non-REM sleep prevent the brain from becoming too cool or too hot. This hypothesis is consistent with the first appearance of cycling sleep in warm-blooded animals but it doesn't explain either the variation in REM across mammalian species or the decrease in REM over the lifespan.

Another idea is that cycling sleep somehow promotes the development of the brain in early life. In particular, cycling sleep may play a special role in the later, mostly postnatal, stages of development that require experience-driven plasticity. The experimental evidence in support of this idea comes from experiments in which kittens have one eye artificially closed for a brief period. This results, within a few hours, in a reduced excitation of neurons in the visual cortex by stimuli (light pulses) delivered to the deprived eye and enhanced responses to stimulation of the open eye. When kittens are allowed to sleep following a period of monocular deprivation, this change in the responsiveness of cortical neurons is retained and even enhanced. But when kittens were either totally sleep-deprived or selectively deprived of non-REM sleep, the effects on cortical neurons of the monocular-deprivation experience were lost. Conversely, in a separate set of experiments, selective deprivation of REM sleep seemed to exaggerate the effects of monocular deprivation, producing even greater changes in the responses of visual cortex neurons.

If cycling sleep were only involved in the experience-dependent phase of brain development, then there would be no need for it to continue into adulthood. One possibility is that it is retained in adulthood but no longer has a function. But this is unlikely. Recall that the cellular mechanisms involved in the experience-dependent phases of later brain development (plasticity expressed as growth of axons and dendrites, and changes in intrinsic excitability and synaptic strength) are retained in the adult brain to store memories. Could the same be true of the sleep cycle? Perhaps alternating periods of REM and non-REM sleep initially serve to consolidate experience-driven changes in late brain development and then remain in a slightly different form to integrate and consolidate memory.

A basic hypothesis of cycling sleep and memory has been nicely articulated by Robert Stickgold of the Harvard Medical School, who writes "the unique physiology of sleep and perhaps even more so, of REM sleep, shifts the brain/mind into an altered state in which it pulls together disparate, often emotionally charged and weakly associated memories into a narrative structure and . . . this process of memory reactivation and association is, in fact, also a process of memory consolidation and integration that enhances our ability to function in the world."

A large number of studies in both humans and rats have shown that a normal night's sleep following certain simple learning tasks results in improved performance when subjects are tested the next day. In most of these studies there is not an absolute requirement for sleep in order to consolidate memory. Some memory for the training experience is still present after 8 hours of wakefulness, and this effect is found whether the wakefulness occurs during the day or at night. But normal cycling sleep produces a noticeable improvement. In a way, these experiments prove something that is widely appreciated in folk traditions around the world: many cultures have a saying to the effect of "sleep on it and you'll have a better understanding of the problem in the morning."

Anecdotal reports of sleep-inspired insight abound. Paul McCartney of the Beatles relates that the tune for the hit song "Yesterday" came to him when he awoke from a dream. The nineteenth-century German chemist Friedrich Kekulé claimed that he solved the ring structure of benzene after being inspired by a dream in which a snake was biting its tail. The American inventor Elias Howe reported that the main innovation allowing for the first sewing machine (placing the thread hole near the tip of the needle) came to him during sleep. But do insight and revelation regularly result from sleep or are these just a coincidences that have resulted in a few good stories?

One interesting study of human learning and sleep deprivation comes from the laboratory of Jan Born at the University of Lübeck in Germany, where investigators sought to test the notion that a night's sleep can help yield insight into a previously intractable problem. To do this, a numerical problem was devised that could be solved by sequential application of simple rules. The experimenters embedded within the problem a shortcut that, if appreciated, could allow the subject to respond much more quickly than through the sequential-application method (see Figure 7.3 for the details of this task). None of the participants recognized the shortcut in the first block of trials. After a night's sleep, though, 13 of 22 subjects had the insight to recognize the shortcut, while, in a different group of subjects, who were not allowed to sleep over a similar interval, only 5 of 22 found the shortcut. The experimenters' conclusion: sleep inspires insight.

A large number of studies have sought to interfere with REM sleep by waking humans or lab animals when an EEG recording indicates that they have entered a REM stage. Selective REM deprivation has been reported to interfere with memory consolidation for a number of learning tasks. In some cases the results have been dramatic: in one report, when humans were trained in a visual texture discrimination task, in which reaction time is taken as a measure of learning, they showed no evidence of learning whatsoever after a REM-deprived sleep but significant learning after either a normal sleep or sleep in which non-REM periods were selectively disturbed. It's important to note that REM deprivation seems to interfere specifically with the consolidation of memories for rules, skills, procedures, and subconscious associations (nondeclarative memory) but not memories of facts and events (declarative memory). Thus the people who spent a REM-deprived night following visual texture discrimination training still had clear memories of the training session (an event) but did not retain their quick reaction times in the task (a nondeclarative skill).

The timing of REM sleep also appears to be important. REM sleep must occur within 24 hours of the training experience in order for it to improve memory consolidation. People who learn a new skill or procedure during the day and then miss that night's sleep will not show any improvement following sleep on the second night. A similar effect is seen in rats, but the interval is reduced: REM sleep must occur within 4–8 hours of training to have a beneficial effect.

REM sleep also appears to be associated with "playback" of the previous day's memories. Kendall Louie and Matt Wilson of MIT used arrays of electrodes to simultaneously record from large numbers of "place cells" (Figure 5.11) in the

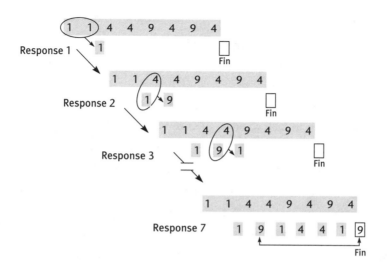

Response 1

Response 2

Response 3

Response 7

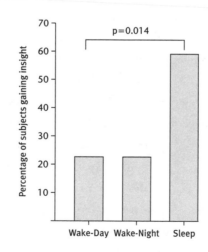

p=0.014

FIGURE 7.3. Sleep as a source of insight. Subjects were trained in a number-reduction task with a hidden rule and then had either intervening sleep, wakefulness during the day, or wakefulness during the night before being retested. The top panel illustrates a sample trial of the task. On each trial, a different string of eight digits was presented. Each string was composed of the digits 1, 4, and 9. For each string, subjects had to determine a digit defined as the "final solution" of the task trial (Fin). This could be achieved by sequentially processing the digits pairwise from left to right according to two simple rules. One, the "same rule," is that the result of two identical digits is just that digit (for example, 1 and 1 results in 1, as in response 1). The other, the "different rule," states that the result of two nonidentical digits is the remaining third digit of this three-digit system (for example, 1 and 4 results in 9, as in response 2). After the first response, comparisons are made between the preceding result and the next digit. The seventh response indicates the final solution, to be confirmed by pressing a separate key. Instructions to the subjects stated that only this final solution was to be communicated and this could be done at any time. It was not mentioned to the subjects that the strings were generated in such a way that the last three responses always mirrored the previous three responses. This implies that in each trial the second response coincided with the final solution (arrow). Subjects who gained insight into this hidden rule abruptly cut short sequential responding by pressing the solution key immediately after the second response. The bottom panel shows the percentage of subjects who gained insight into the hidden rule following sleep versus two conditions of wakefulness. Reproduced with permission of Macmillan Publishers, Ltd., from U. Wagner, S. Gais, H. Haider, R. Verleger, and J. Born, Sleep inspires insight, *Nature* 427:352–355 (2004).

hippocampus of rats as they repeatedly ran a unidirectional path in a circular track to obtain a food reward. The experimenters were able to see sequential activation of place cells coding for various locations on the circular track as the animal ran. Then recordings were continued as the animal slept after training. Amazingly, these same patterns of hippocampal place cell activation were replayed during REM sleep. The replay wasn't a perfect spike-for-spike reproduction of the waking activity. Sometimes the pattern was a bit degraded and sometimes the pattern was recognizable from the waking experience, but the overall speed of the activity had changed. Nonetheless, this study, and several others like it from different laboratories, have found statistically significant reactivation of neuronal ensemble activity during REM sleep following training. Was the replay of activity in Louie and Wilson's rats important for consolidating memory of the circular track? If so, what aspects of the experience? Were the rats dreaming of the circular track when the replay activity was recorded during REM sleep? We don't yet know the answer to these questions.

One might be tempted to conclude from this line of evidence that the relationship between REM sleep and memory consolidation is fairly solid. But a bit more investigation will reveal some cracks in the façade. For example, subsequent experiments on both rats and humans have shown that selective deprivation of non-REM sleep can also have deleterious effects on consolidation of some nondeclarative memory tasks, although these tend to be smaller than those achieved by selective REM sleep deprivation. In addition, a recent report indicates that the "playback" of neuronal firing patterns following novel experience in the rat is actually stronger in deep non-REM sleep (stages III and IV) than it is in REM sleep. Most important, it is almost impossible to produce REM sleep deprivation without also causing stress and the accompanying rise in circulating stress hormones. We know that stress can impair learning in both humans and rats and that both stress and artificial administration of stress hormones can interfere with synaptic and morphological plasticity in rat brains.

Finally, there is a strong prediction of the REM sleep and memory consolidation hypothesis that has not been born out. Modern antidepressant drugs, including the serotonin-specific reuptake inhibitors (SSRIs, such as Prozac and its kin) and tricyclic antidepressants (such as Elavil), produce a partial reduction of REM sleep. But an earlier class of antidepressants, the monoamine oxidase inhibitors, such as phenelzine (Nardil), produce a complete blockade of REM sleep. A similar effect is seen with certain forms of traumatic brainstem damage, yet both of these cases that produce complete blockade of REM sleep (and do so without stress hormone surges) do not seem to produce significant impairment of memory. Conversely, the benzodiazepine class of anti-anxiety drugs (including Valium, Xanax, and Versed) have strong memory-blocking effects, yet leave sleep cycles unperturbed.

So, what are we to conclude? The evidence that cycling sleep has some role in the consolidation and integration of memory is fairly good. The notion that REM sleep has a privileged part in this process is somewhat weaker. My own guess is that a holistic explanation is more accurate: it's likely that something about the cycling between REM and non-REM stages throughout the night is particularly beneficial in memory consolidation and integration. Some theoretical models, involving alternating unidirectional flow of information between the hippocampus and the cerebral cortex, suggest why this might be, but I won't go into those details (interested readers are encouraged to check the Further Reading and Resources section).

So what's special about sleep? Perhaps the type of integration and cross-referencing that sleep allows is somehow different than that of the waking state. One might imagine that the reduction of external sensation during sleep

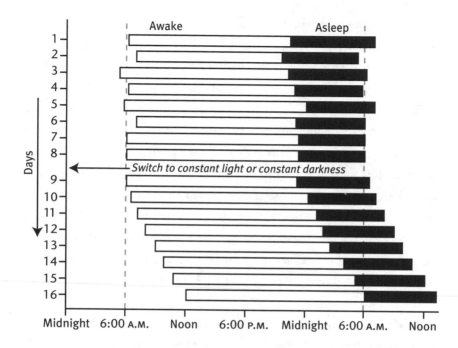

FIGURE 7.4. Changes in the human daily sleep-wake cycle in the absence of external cues. The cycle persists in the absence of cues from alternating light and darkness, but becomes gradually desynchronized to the external world. In this diagram hollow bars represent waking and filled bars indicate sleep. *Joan M. K. Tycko, Illustrator.*

allows for associations between more distant and fluid aspects of memory that would be impossible during waking sensory bombardment. Let's keep this thought in the back of our minds and return to it shortly when we consider dreams.

TO THIS POINT, I have discussed the sleep-wake cycle and the stages of sleep without reference to the brain circuitry and molecular events underlying them.

Let's move in that direction by asking a very fundamental question: do daily cycles of activity such as the sleep-wake cycle require a sort of clock within the brain, or is this behavioral rhythm solely driven by external cues, such as those from sunlight? Figure 7.4 shows what happens when someone who has been living in normal conditions for 10 days, with light and dark cues, is placed into conditions where these cues are no longer present (either constant light or constant darkness). The basic daily rhythm of sleeping and waking persists with a near 24-hour-long cycle (about 24.2 hours on average), but this cycle becomes gradually desynchronized from the clock of the external world, and the time of sleep onset slowly shifts later and later. This indicates that there is indeed a clock within the brain but that it requires information to remain synchronized to the outside world.

It turns out that a tiny structure within the hypothalamus called the suprachiasmatic nucleus (that means "above the place where the optic nerves cross" and is abbreviated SCN) is the body's master timekeeper. This cluster of about 20,000 neurons has a natural rhythm of activity that continues even if you remove it surgically (from a hamster, for example) and grow it in a lab dish filled with nutrient fluids. This activity is approximately, but not exactly, 24 hours long, hence its name, the circadian clock (from circa = approximately and dia = day). Animals that sustain damage to the SCN no longer have normal sleep-wake cycles. Rather, they have brief periods of sleep and waking distributed randomly throughout the day and night.

The way light coordinates the timing of the internal circadian clock with the external world is mostly driven by a special set of neurons in the retina. These are not the rods and cones that form the visual image, but rather a group of large, spindly cells called melanopsin-positive ganglion cells. These cells send their axons to the SCN to give information about the ambient light level. Significantly, not only are melanopsin-positive ganglion cells stimulated by strong

FIGURE 7.5. A rendition of Carl von Linné's flower clock, which uses the opening and closing times of European flowers to estimate time of day. *Joan M. K. Tycko, illustrator.*

sunlight, but they can also be activated by relatively weak artificial lighting. Therefore, when you stay up late under artificial light you are trying to force your internal circadian clock into a 25- or 26-hour period. The result: morning grogginess. The degree to which light can shift the internal circadian clock is limited to about a 1-hour shift per day. So, if you make a flight across 5 time zones you are likely to need about 5 days for your internal clock to reset to the new local time. The result, as you well know: jet lag.

Is the circadian clock solely a device to drive the sleep-wake cycle? After all, many organisms have functions that are coordinated to the time of day but are independent of sleeping. Even many plants open or close their flowers at particular times of the day (Figure 7.5). This was noted by the Roman philosopher Pliny the Elder, writing in the first century A.D., and was elaborated by the eighteenth-century Swedish naturalist Carl von Linné, who proposed that it would be possible to create an accurate clock by planting a flower garden with carefully calibrated opening and closing times. It turns out that the basic biochemical scheme of the circadian clock found in the human SCN can also be found in lower animals, plants, and even fungi. Clearly, the ability to coordinate biological processes with the light-dark cycle is an important function that is likely to have predated sleeping animals by a billion years. It's most likely that circadian clocks evolved independently, at least twice: fungi have circadian clock genes that are related to ours, but cyanobacteria (as well as archaea and proteobacteria) have a set of unrelated molecules that nonetheless perform similar functions. Interestingly, these ancient bacteria are likely to have developed their circadian clock about 3.5 billion years ago when the Earth's rotation period was only about 15 hours (this is an estimate).

What is it that originally drove the evolution of the circadian clock? We don't know the answer to this question and several hypotheses have been put forward. One appealing idea, formulated by Colin Pittendrigh in the 1960s, is called the "escape from light" hypothesis. Pittendrigh and others noticed that several species of unicellular algae underwent replication of their DNA and subsequent cell division only during the night. It was known that dividing cells can be killed by the ultraviolet radiation present in daylight. Hence, Pittendrigh suggested that circadian rhythms evolved as an escape from light: to allow sensitive cellular processes to occur in darkness. Recently, Selene Nikaido and Carl Johnson of Vanderbilt University put this to the test: They showed that the unicellular alga *Chlamydomonas reinhardtii* survives exposure to a pulse of ultraviolet light best during the day, when cell division ceases. When lab dishes of *Chlamydomonas* were placed in constant light conditions, they had a persistent circadian cycle of cell division that gradually became desynchronized with the outside world, just like the sleep-wake cycle of humans kept in constant light.

ALTHOUGH RECENT YEARS have seen an explosion of knowledge about the molecular basis of the circadian clock in the SCN and about the brain circuitry involved in the onset and various stages of sleep, the way in which the SCN affects sleep-control circuits is still poorly understood. Axons from the neurons of the SCN make synapses in several adjacent regions of the hypothalamus that in turn project to brainstem and thalamic structures. In addition, the SCN, through a complex circuit of at least three synaptic relays, stimulates the pineal gland to secrete the hormone melatonin. The levels of melatonin, widely sold in health food stores as a sort of "natural sleeping pill," increase with nightfall and peak at about 3:00 A.M. Melatonin diffuses throughout the body, but has its major effect on sleep-control circuits in the brainstem.

One of the main circuits in the brain that affects the control of sleep is called the brainstem reticular activating system. These neurons, which use the transmitter acetylcholine (and are hence called cholinergic neurons), send their axons to sites in the thalamus, where they modulate the transmission of information between the thalamus and the cortex. Reticular cholinergic neurons are active during waking but gradually become less and less active as non-REM sleep progresses to deeper stages. Indeed, artificial electrical stimulation of the reticular activating system will wake an animal from sleep while stimulation of its targets in the thalamus will have the opposite effect: it will induce deep non-REM sleep in a previously awake animal. When the transition from non-REM to REM sleep begins, the brainstem cholinergic neurons begin firing rapidly again, and this causes the EEG record to shift from the large-amplitude, synchronized state to the small-amplitude, desynchronized state that's typical of both REM sleep and waking. Why doesn't the animal just wake up at this point, instead of staying in REM sleep? The answer is that other brainstem systems, the serotonin-containing neurons of the dorsal raphe and the noradrenaline-containing neurons of the locus coeruleus are also involved in sleep-cycle control, and these neurons are inactive in both REM and non-REM sleep. The interaction of these three brain regions (together with some others that play a smaller role) determines how sleep stages progress through the night. The large number of neurotransmitter systems involved in sleep-cycle control means that a variety of drugs can affect sleep, producing either a desired effect (such as sleep through the use of drugs that interfere with acetylcholine receptors) or an unwanted side effect (such as the REM sleep–inhibiting properties of many antidepressants that boost serotonin).

EVERYONE LIKES TO talk about dreams. The thing about dreams is that they feel inherently meaningful. In every culture studied to date, people have elaborate ideas about the meaning and causes of dreams. In many cases, dreams are thought to be messages from divine beings or ancestors that can provide guidance or foretell the future—the Judeo-Christian Bible, the Islamic Koran, and the sacred texts of Buddhism and Hinduism all contain stories of prophetic dreams. Dreams can also be thought to represent "soul-travel" to distant locations. If you believe that dreams are meaningful, you can hold that their meanings are either rather straightforward, reflecting prior events and concerns, or are occluded and symbolic, requiring interpretation. The ancient Egyptians, by about 1500 B.C., had elaborate temples that were specifically built for dream interpretation by trained priests. Manuscripts survive from this time that catalogue the meanings of various dream elements. Most of these are couched in terms of prophecy ("if you dream of crows, then a death will soon come to a loved one").

Many years later, Sigmund Freud, the father of psychoanalysis, would elaborate a related theory in his famous 1900 volume entitled *The Interpretation of Dreams*. In Freud's view, dreams arise from subconscious wishes, mostly of a sexual or aggressive nature, that the conscious mind suppresses during the day. But if these subconscious wishes were manifest in dreams in a straightforward fashion, then the dreamer would be awakened by these forbidden desires. So, instead, dreams are symbolic reflections of the dreamer's suppressed subconscious wishes. Thus, in Freud's view, a dream of flying represents displaced sexual desire, and a man's dream of teeth falling out represents a fear of castration (it's unclear what such a dream would mean for a woman). In many ways the practices of ancient Egyptian dream priests and those of present day post-Freudian psychoanalysts are not dissimilar. They have different goals in that the former are concerned with predicting the future while the latter seek to illuminate past and present events and motivations. But both rely, more or less, upon a dictionary of symbols to guide dream interpretation.

There is no question that dreams *feel* meaningful and symbolic. Indeed, various symbolic dictionaries for dream interpretation (in the basic form of "if you dream of X then it means Y") are sold by the tens of thousands every year. Although dream interpretation is a phenomenon that is broadly cross-cultural, it is not accepted by all. There are those, mostly a subset of neurobiologists, who hold that the content of dreams has no meaning whatsoever. In their view dreams are merely the byproduct of some other important process, such as memory consolidation. Dreams are the smoke and not the fire, so to speak.

Let's try our best to address this contentious issue systematically. First, we'll consider some ideas about how patterns of activity in the brain might give rise to dreams. Then, we'll talk about the possible function or purpose of dreaming, and finally we'll attempt to ask whether the content of dreams is meaningful.

You know from your own experience that some mornings you may awaken with no recollection of any dreams at all, while at other times the night seems to be crowded with them. In general, unless you awaken during or within a few minutes of the end of a dream, you are unlikely to recall it. For many years it was thought that dreaming only occurred during REM sleep. Now we know that dreams can be reported following awakening from any stage of sleep but that their character, duration, and frequency vary with different sleep stages. Let's illustrate this with some examples from my own dream journal.

Dream 1: Shortly after falling asleep, I had the sensation of swimming underwater, as I did with my kids at the neighborhood pool yesterday.

Dream 2: I couldn't get anything done on my grant application today, and was plagued through the night with worry that I couldn't finish it before the deadline.

Dream 3: I am waltzing with a beautiful woman in a vast space. The woman is not someone I recognize but she seems to know me well. In some respects the room where we're dancing is like a large ballroom, but it's also like a shop in my home town that I visited frequently as a teenager. This shop sold musical instruments, including many unusual ones from foreign countries. My dancing partner is beaming at me, but I'm distracted by the instruments in the cases, which are complex and inviting. I long to go tinker with them, but I'm aware that my dancing partner is getting annoyed that I'm not paying enough attention to her. She grows more and more upset as she senses my distraction. Soon, she's furious and I'm running from her and the scene has changed to a long, hot road. I jump on a bicycle and pedal quickly, which allows me to pull away from her pursuit. I can no longer see her in the road behind me. However, after a minute or so, the road grows bumpy and I realize that I'm riding over live snakes. As I pedal, the snakes snap at my feet each time they reach the lowest point in the pedal's revolution, so I put my feet up on the crossbar of the bike to avoid being bitten. Of course, I gradually lose speed and I realize that very soon, without forward momentum, I will lose my balance and fall into the snakes that now cover the road like a carpet.

Heaven knows what a psychoanalyst (such as my father!) would make of all this (is a snake, sometimes, just a snake?). These dreams are very different, but they do share two common features: I am the main character and they occur in the present. This is a general feature: the vast majority of dreams are "present-tense, first-person" experiences. Dream 1 is a typical dream from the period shortly after sleep onset. It is brief, and while it has a strong sensory component, this does not progress to form a continuing narrative. It is a scene fragment without much detail and without any particular emotional tone. It is logical, congruent with waking experience, and does not have hallucinatory properties. Significantly, sleep-onset dreams are very likely to incorporate experiences from the previous day's events. In one study, Robert Stickgold and his coworkers from Harvard Medical School had subjects play the video game "Downhill Racer II" for several hours. In the following night's sleep, more than 90 percent of the subjects reported scenes from this game, but only when they were awakened shortly after sleep onset, not in the middle or late parts of the night when deep non-REM (stages III-IV) and REM sleep predominate.

Dream 2 is a typical dream from deeper, non-REM sleep, particularly as would be found in the first half of the night. Like Dream 1, it lacks an unfolding story, but in this case, it almost completely lacks sensory experience. Basically, it's just an obsessive, emotion-laden anxious thought. The thought is logical and grounded in waking experience, but it does not trigger any form of narrative.

Dream 3 is typical of REM sleep, particularly REM episodes that occur shortly before waking. It is a narrative dream that unfolds in a story-like fashion and is rich in detail. The dream fuses together disparate locations, some specific (the music store of my youth) and others generic (a fancy ballroom I don't recognize). It incorporates elements of fantasy: in real life, I can't waltz for beans, but in the dream I do it flawlessly and without effort. There is a sense of continuous motion throughout the dream (waltzing, running, cycling). The dream narrative incorporates scene changes (from the ballroom to the road) and other events and locations that don't make sense, and yet, in the dream, I accept these phenomena as the natural course of things. There is a suspension of disbelief about otherwise illogical or bizarre experiences. There are many hallucinatory aspects to this dream but they are almost exclusively visual (as opposed to auditory or tactile). Finally, there is a growing sense of anxiety and fear that builds throughout the dream, starting with the mild social anxiety of offending my dancing partner and culminating with the acute fear of a horrible death by snakebite.

Narrative, emotion-laden dreams with illogical and bizarre scenes are the kinds of dreams we are most likely to remember and discuss, partly because they make for good stories, but also because of the structure of the sleep cycle: you are most likely to awaken, and therefore remember your dream, toward the end of the night's sleep when REM predominates. This type of dream is most frequent during REM, but we have recent evidence that people awakened from non-REM sleep during the last third of the night can sometimes recall similar narrative dreams.

There have now been many large studies in which people have kept dream journals (either written or audio) and a much smaller sample in which people in a sleep lab or wearing a home EEG recording unit are awakened during various sleep stages to provide dream reports. What becomes clear from these studies is that, in general, dream content is very highly biased toward negative emotional states. Fear, anxiety, and aggression are the dominant emotions in about 70 percent of dreams recorded in dream journals. Only about 15 percent of these dreams are clearly emotionally positive. These results seem generally to hold cross-culturally: dreams of being chased are the most common single theme found around the world, from Amazonian hunter-gatherers to urban dwellers in Europe. Interestingly, the proportion of dreams with prominent anxiety, fear, and aggression is greater in dream journals that rely upon spontaneous waking than it is in situations where people are awakened artificially in the last third of the night (reduced from 70 percent to about 50 percent). One interpretation of this disparity is that dreams with negative emotions are more likely to awaken the sleeper, who will then remember and record them.

Given the preponderance of sexual dream interpretations by Freud, it is interesting that less than 10 percent of dreams appear to have overtly sexual content. This is similar in men and women. The previously mentioned male and female genital responses that occur during REM sleep do not seem to be correlated with sexual dreaming.

Elements of the previous day's activity, particularly those with a strong sensorimotor component, are often incorporated into brief sleep-onset dreams, but seldom in narrative dreams. In one study, less than 2 percent of narrative dreams contained autobiographical memory replay of an event from the previous day (although more incorporated a single aspect of the day's experience such as a person or a location). Some researchers have claimed that there is a time-lag effect in which experiences are most likely to appear in dreams 3 to 7 nights later. Counterintuitively, highly emotional experiences during the day seem to require a slightly longer time before showing up in dreams.

So, let's summarize the differences between the waking state and narrative dreaming. Compared to the waking state, narrative dreaming

incorporates bizarre aspects, including fusions and abrupt changes of locations and individuals, violation of physical laws, and so on;

is characterized by a lack of internal reflection and an acceptance of illogical events;

often involves a heightened sense of motion, predominantly conveyed visually;

has a higher incidence of negative emotion than waking life, particularly anxiety and fear;

incorporates older memories to a greater degree than new ones;

is rapidly forgotten unless interrupted by waking.

In recent years, a number of studies have used scanners to measure brain activity in people during various stages of sleep. Let's examine these findings with an eye to whether they can help explain some of the characteristics of narrative dreaming listed above. Although narrative dreams can occur during either deep non-REM sleep or REM sleep, they seem to predominate in the latter, so we'll use the REM sleep stage brain as our template for a physiological analysis of narrative dreaming. Figure 7.6 shows a simplified summary of changes in brain activity during REM sleep as compared to restful waking.

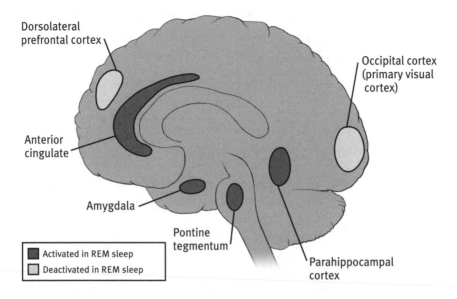

Dorsolateral
prefrontal cortex

Occipital cortex
(primary visual
cortex)

Anterior
cingulate

Amygdala

Pontine
tegmentum

Parahippocampal
cortex

■ Activated in REM sleep
□ Deactivated in REM sleep

FIGURE 7.6. Some brain regions that show altered electrical function in REM sleep, as determined by PET scans. This figure is not meant to be complete. For example, in addition to the amygdala and anterior cingulate, adjacent parts of emotional circuitry are also activated during REM sleep, including the septal area and infralimbic cortex. Adapted from J. A. Hobson and E. F. Pace-Schott, The cognitive neuroscience of sleep: neuronal systems, consciousness, and learning, *Nature Reviews Neuroscience* 3:679–693 (2002). *Joan M. K. Tycko, illustrator.*

We knew from previous work in animal models that the brainstem reticular activating system is strongly active during REM sleep, and the activity of these cholinergic neurons (in a place called the pontine tegmentum) can be seen in PET scan images. One of the most striking features of the brain scans is that while narrative dreams are intensely visual, the primary visual cortex is almost completely silent during REM sleep. But areas involved in the higher-level analysis of visual scenes and the storage of visual and cross-modal memories (such as the parahippocampal cortex) are strongly activated. This may help explain why dreams are often constructed from fragments of disparate memories, mostly long-term visual memories stored in these visual association areas.

Another striking feature of the brain in REM sleep is strong activation of regions subserving emotion. In particular, the amygdala and anterior cingulate are strongly activated and these regions appear to play a particular role in fear, anxiety, and the emotional aspects of pain as well as responses to fearful and painful stimuli. This may underlie the prevalence of fear, anxiety, and aggression in the emotional tone of narrative dreams. Finally, portions of the prefrontal cortex, in particular the dorsolateral prefrontal cortex, are deactivated in REM sleep. This is a crucial part of the brain for executive functions (judgment, logic, planning) and working memory. Its deactivation may help explain the illogical character of dreams and the dreamer's acceptance of bizarre and improbable circumstances and plotlines. Essentially, reduced dorsolateral prefrontal activation could contribute strongly to the hallucinatory properties of dreams. In this sense, it is worth mentioning that deactivation of this region is a hallmark of hallucinating schizophrenics (who, in a limited sense, have dreamlike experiences while awake).

Brain scanning with PET is a technique that gives information about the average activation of brain regions. It is very useful, but it does not convey detailed information about either the exact location of individual firing neurons or the fine temporal structure of that activity. These parameters are both critical to understanding the way information is being processed in the brain during narrative dreaming. Animal experiments using implanted recording electrodes have shown that during REM sleep the noradrenaline-containing neurons of the locus coeruleus and the serotonin-containing neurons of the dorsal raphe fall silent, while the acetylcholine neurons of the brainstem reticular activating system fire strongly. The neurons of these three modulatory systems have axons that project widely throughout the brain, including the thalamus, limbic system, and cortex. Thus some of the regional activity changes during REM sleep, as reflected in brain scanning studies, results from turning up synaptic drive that uses acetylcholine together with turning down that which uses noradrenaline and serotonin.

The increased cholinergic drive also ultimately results in the limp muscle paralysis that characterizes REM sleep. During narrative dreams, the motor cortex and other movement control structures such as the basal ganglia and cerebellum are issuing commands to cause movements, but these commands are blocked from entering the spinal cord by an inhibitory circuit triggered by strong cholinergic drive in the brainstem. This may underlie the continual and effortless sense of movement (including flight) that is so prevalent in the experience of narrative dreams: the commands for movement are being issued but the feedback from the muscles and other sensory organs about how those movements are progressing is no longer present to ground the perception of movement in reality.

ALTHOUGH THE STORY is far from complete, we can certainly say that the pattern of brain activity during narrative dreaming can explain many of the unusual features of dream content. This level of explanation, does not, however, address either the purpose of dreams or the question of whether dream content is meaningful. So, why do we dream? The short answer, sadly, is that we don't really know. The long answer, however, suggests some avenues for investigation.

If you ask a cross section of sleep researchers why we dream, you tend to get answers that reflect that person's area of interest. In this fashion, scientists whose primary interest is emotion will tell you that the main function of dreams is to regulate mood. For example, Rosalind Cartwright, of Rush Presbyterian–St. Luke's Medical Center in Chicago, theorizes that dreams function as mood regulators, to allow us to process negative emotions, so we wake up feeling better than we were when we went to sleep. Some psychiatrists say that dreams are like a kind of psychotherapy. Ernest Hartmann of Tufts University has proposed that both dreams and psychotherapy largely function by allowing connections between life events to be made in a safe, insulated environment, away from the outside world.

Biologists with an interest in evolution have proposed that dreaming has developed as a time to rehearse and perfect behaviors that are crucial to survival during waking hours. They function as a kind of virtual-reality environment to simulate life-threatening scenarios in a safe place. In a way, this explanation is not too different from that offered by Hartmann. Both seek to explain the central role of fear and anxiety in dream reports, and both imagine dreaming as a protected environment in which to accomplish important mental tasks.

And, of course, I've already discussed the idea that cycling sleep is important for the consolidation, integration, and cross-referencing of memory, so it is a small leap to imagine that dreaming is somehow related to these memory processes. One interesting twist on this comes from Jonathan Winson of Rockefeller University, who thinks of dreams as "off-line memory processing." In his view, the computational resources needed to integrate experience into memory, if operative only during waking, would require an even larger and ultimately untenable volume of cortex than we already have. So, in order to make the best of the brain volume we have, we run the night shift, so to speak, continuing the process of memory consolidation and integration around the clock, like a wartime munitions factory.

In considering the merits of these models for dream function we should keep several things in mind. First, these models are not necessarily mutually exclusive: for example, dreams could function *both* as regulators of mood and as a part of memory consolidation. Second, we need to be careful to make some important distinctions between levels of analysis in dreams. On one level are the underlying processes that occur in the brain during the dreaming state. Then there is the experience of the dreaming state while it is happening, and, finally, the report of the dream that will only occur for those dreams that are interrupted or that are followed very quickly by waking.

In my view, each of the models for dream function has some strengths and weaknesses. The psychiatric explanations of dreams as mood regulators or as night therapy provides a plausible rationale for the prevalence of negative emotion in dream reports. But this model has to contend with two important observations. First, there are some people who report no dreams at all unless artificially awakened, and yet, on the average, these people have no unusual incidence of emotional or cognitive problems. Here, one might retreat a bit and posit that the therapeutic value of dreams occurs as a result of their experience during sleep, even if they are not consciously recalled. Second, many of the most emotionally salient events in life never make their way into dreams at all, even in those individuals who report dreams regularly. Some psychiatrists might respond that these events would be manifest symbolically rather than literally and would therefore not always be easy to spot.

The memory consolidation/integration model for dreams is compelling in many ways. Among other things, it provides an explanation of why items in remote memory are often dredged up in dreams: presumably these are being integrated with newer memories. Within memory consolidation/integration models there are some important distinctions. In some, the experience and/or later report of the dream are central to the process. These models, of course, must deal with the same critique leveled at the emotional models above: on average, people who fail to report dreams perform normally in a battery of memory tests. A reductionist variant of the memory model, most forcefully proposed by

J. Allen Hobson of Harvard University, states that the main purpose of cycling sleep is memory consolidation and integration and that the experiences of narrative dreams are basically what the logically impaired (inhibited dorsolateral frontal cortex) and hyperemotional (overactive amygdala, septum, and anterior cingulate) brain can stitch together into a narrative from scraps of mostly visual memory (overactive parahippocampal gyrus). In this view, the content of dreams is merely a funhouse-mirror reflection of memory consolidation and there is no need for symbolic dream interpretation in the Freudian (or ancient Egyptian) tradition.

To me, there has always been a big hole in existing memory consolidation/integration models of dreams. They fail to address why the emotional content of dreams is so negative. My own suspicion about this has been as follows: It is well known that the activation of the negative emotion circuits (fear/anxiety/aggression) in the brain will reinforce memory consolidation in the waking state. Essentially, strong activation of the brain regions subserving negative emotions is a signal that says "write this down in memory and underline it." During memory consolidation and integration in sleep we need some mechanism to say, "OK. You've made this connection with something in long-term memory. Write it down now." I suggest that that mechanism is activation of the negative emotion centers. In essence, the fear/anxiety/aggression circuitry is co-opted for use in reinforcing memories and connections between memories in the absence of relevant emotional stimuli. Your dreaming brain doesn't know that the negative emotion circuits have been hijacked, and it integrates the activity in these centers to produce narrative dreams with negative emotional themes.

So, where do these various models leave the question of whether dream content is meaningful? To me, this has always seemed like a nonissue. Certainly, the content of dreams is of some interest under any model of dreaming. Even diehard proponents of the memory consolidation/integration model of dreaming agree that the content of what is being written into memory and what it's being integrated with are of some value in understanding an individual's mental state. The question is how far to take it. Although there is a place for the analysis of dream content in both psychotherapy and personal growth, I have no confidence (and there is no biological basis to believe) that insight into one's mental state can be gained through analysis of dream content with arbitrary symbolic dictionaries.

The obsession with specific dream content tends to obscure what's really important about dreaming. The most useful thing about the *experience* of dreaming (as opposed to the underlying processes) is not the detailed content of dreams. It's not so crucial that you dream of a cigar rather than a shoe, or of your father rather than your mother. What's most important about dreaming is that it allows you to experience a world where the normal waking rules don't apply, where causality and rational thought and our core cognitive schemas (people don't transform or merge, places should be constant, gravity always operates, and so forth) melt away in the face of bizarre and illogical stories. And, while you dream, you accept these stories as they unfold. Essentially, the experience of narrative dreams allows you to imagine explanations and structures that exist outside of your waking perception of the natural world. In your waking life you may embrace the distorted structures of the dream world or you may be a hard-headed rationalist, or you may blend the two (as most of us do), but in all cases the experience of dreaming has thrown back the curtain and allowed you to imagine a world where fundamentally different rules apply.

Chapter Eight

The Religious Impulse

IN MY NIGHTMARE, I'm in New Orleans for the Society for Neuroscience annual meeting, a gathering of 30,000 or so of the world's brain researchers. It's nighttime and I'm at a restaurant table with a group of colleagues. The wine is flowing, everyone is happy and chatting, so I begin to explain my theory of religion and neural function as the waiter delivers the huge plates of steaming boiled crawfish. As I go on for a minute or two I slowly realize that the table has become oddly silent. Behind me, a tall robed figure with a black hood is waiting expectantly with a pepper grinder about the size of a Stinger missile. I turn slowly, my spiel gradually winding down.

"Would you care for some freshly ground speculation with that, sir?"

All heads turn toward me. The sounds in the restaurant gradually build from a low rumble to howling to shrieking cacophonous laughter. Then the hooting begins and all the diners slowly point their fingers in my direction at once. The gathering din soon animates the boiled crawfish on my plate and they jerk and snap and finally swarm over me, ripping at my flesh and singing "In-a-gadda-da-vida baby, don't you know I'm in love with you" in tinny little voices as I crumple to the floor.

I'm not alone. Neurobiologists are hesitant to talk about brain function and religion in the same breath. Every human culture has language and music, and we are happy to study the neurobiological bases of these phenomena. Every human culture has a form of marriage, and we study the neurobiological basis of pair bonding as well. Every human culture has religion. The forms of religion vary enormously (as do languages and marriage customs), but the presence of religion is a cross-cultural universal. To date, a culture has yet to be found that lacks religious ideas and practice. Yet scientists who study the brain rarely contribute to discussions of this widespread form of human behavior (perhaps out of a widespread fear of singing zombie crawfish from hell).

FRESHLY GROUND SPECULATION in hand, let's have at it and consider some religious ideas from around the world. Some of these come from a provocative book by the cognitive anthropologist Pascal Boyer, *Religion Explained.*

> Invisible souls of dead people lurk everywhere. They must be pacified with offerings of food and drink or they will make you sick.

> There has only been one woman in history who has given birth without having sex and we worship her for that reason.

After you die, you will come back to earth in either a higher or lower form, depending upon how you have followed a set of rules in this life.

There is one god who is all-powerful and all-knowing and who can hear your thoughts. You can pray to our god in a temple or anywhere else.

Some ebony trees can recall conversations people hold in their shade. These can be revealed by burning a stick from the tree and interpreting the pattern of fallen ash.

There is a shaman in our village who will dance until his soul leaves his body and goes to the land of the dead. When he returns he will bring messages from our ancestors who have become all-seeing gods.

It is likely that some of these ideas are from traditions with which you are familiar and others are not. This small collection illustrates a bit of the variety of cross-cultural religious thought. Some groups have religions with one god, others have many, and some have none at all. In some cases, unusual powers are attributed to historical figures or natural objects, which then become the focus of particular attention. In others, special rituals can be used to speak with divine beings or the dead.

There is a lot of variety here, but not infinite variety. You don't, for example, find a religion where there is an all-powerful, all-seeing god but he never interacts with the human world, or one in which the spirits of the ancestors will punish you for doing what they want, or one in which priests can see the future but then forget what they saw before they can tell anyone. Religions, like dreams, have variety, but they are still constrained within a particular set of cognitive and narrative boundaries.

So, why is it that religion of some form is found in every culture (although not in every individual)? Why is it that, in Boyer's words, "human beings can easily acquire a certain range of religious notions and pass them on to others?" Can our present knowledge of brain function provide any form of explanation for the prevalence and practices of religion cross-culturally?

If, after a few pints, I start to ask people at my local bar about the origin of religious thought around the world, I get the sort of answers that can be summarized as follows:

> Religion provides comfort, particularly in allowing people to face their own mortality.
>
> Religion allows for the upholding of a particular social order: it lays down moral rules for interactions with others.
>
> Religion gives answers to difficult questions such as what are the origins of the natural world.

These ideas all hold true to some degree for most of the religions we encounter in the more affluent parts of the world. But they do not always apply in the broader cross-cultural sense. Many religions are not comforting at all: they are mostly concerned with malevolent spirits that, if not continually appeased, will kill you, make you sick, make you crazy, destroy your crops, cause you to fail at hunting. Most religions have a world origin story and an afterlife story, but these are not universal. Religions do not always promise salvation. In many of the world's cultures, the dead are condemned to wander eternally no matter how scrupulously they lived their lives on Earth. Many societies have common rules for social order, but in many cases these are entirely independent of religious practice. In short, the explanations offered at the bar have some utility, but they all fail the broader cross-cultural test. They do not answer our basic question: "Why does every human culture have religion?" A different approach is needed.

IS IT REASONABLE to imagine that brain function, something that is generally shared by human beings around the world, can be invoked to explain religious thought and practice, which takes such a wide variety of forms (including atheism)? Let's be clear what we're after here. We're not looking for a brain region or neurotransmitter or gene that somehow confers religion. That is unlikely to be a fruitful level of analysis. Nor are we seeking to explain specific religious ideas in biological terms. Rather let's ask: are there some aspects of brain function that, *on the average,* make it easy for humans to acquire and transmit religious thought?

I will try to convince you that our brains have become particularly adapted to creating coherent, gap-free stories and that this propensity for narrative creation is part of what predisposes humans to religious thought. Creating a coherent percept from sensory fragments is a theme touched on in Chapter 4. Recall that as you scan a visual scene with tiny jumps of your eyes, called saccades, your brain plays some tricks. You do not see a jerky image with the visual scene jumping around, nor do you see a scene that briefly fades to black every time your eyes jump. Rather, your brain takes the "jerky movie" that is the raw visual feed from your eyeballs, edits out the saccades, and retroactively fills in the gaps in the ongoing visual scene with images from the time the saccade ended. What you perceive feels continuous and flowing, but it is actually a narrative actively constructed by your brain to create a coherent sensory story.

The creation of coherent narratives in the brain is not limited to manipulation of low-level perception, as occurs with visual saccades, but extends to higher perceptual and cognitive levels. This function is ongoing, but difficult to study in the normal brain. It is often more clearly revealed in cases of brain damage. Take, for example, people suffering from anterograde amnesia. Recall that these people are unable to form new memories of facts and events but have intact memories for things in the more distant past. When a hospital-bound man with severe anterograde amnesia is asked "What did you do yesterday?" he does not have any memories from the previous day to call to mind. In many cases, the patient will construct a narrative from scraps of older memory and weave them together to make a coherent and detailed story. "I stopped in to visit my old pal Ned at his store and then we went out for lunch at the deli. I had a corned beef sandwich and a dill pickle. Afterward we took a walk in the park and watched the skaters." This process, which is called confabulation, is not merely a face-saving attempt. In almost all cases, amnesiacs believe their own confabulations and will act upon them as if they were true. Confabulation in anterograde amnesia is not a process under voluntary control. Rather, it's what the brain does when confronted with a problem it cannot begin to solve: it makes a story from whatever bits of experience it can dredge up, in much the same way that narrative dreams are created from scraps of memory.

The drive to create coherent narrative is also revealed in a fascinating group of "split-brain" patients. These are people whose severe and otherwise intractable epilepsy has been controlled by cutting the corpus callosum and the anterior commissure, which are the bundles of axons that normally connect the right and left cerebral hemispheres. The split-brain operation, though used as a last resort, is a remarkably effective way of controlling some types of seizures. This procedure disconnects the direct communication between the right and left cerebral cortex, but each side of the cortex retains generally normal function and the lower (subcortical) parts of the brain remain connected. Remarkably, if you meet someone who has had split-brain surgery you are unlikely to notice anything amiss in casual conversation. It takes careful examination, usually employing special instruments, to reveal anything unusual.

The analysis of perception and cognition in split-brain patients was pioneered in the 1960s by Roger Sperry of the California Institue of Technology (the same remarkable neurobiologist whose work on the development of the visual system of the frog I discussed in Chapter 3) and has been carried on more recently by a number of others, including Michael Gazzaniga of the University of California at Santa Barbara. In most people (almost all right-handers and about half of lefties), the left cortex is specialized for abstract thought, language (particularly involving the meanings of words), and sequential mathematical calculation. The right cortex excels at spatial relationships, geometry, face recognition, and detecting the emotional tone of language, music, and facial expressions. These insights have mostly come from patients with various localized brain lesions as well as from brain-scanning studies of normal humans.

Split-brain patients provide a unique opportunity to see how the left and right cortices process information independently. In one famous experiment, a split-brain patient was placed before a specially constructed screen, designed so that the left cortex received only an image of a chicken's claw (projected in the right visual field: the representation of the visual field is reversed right to left in the brain) while the right cortex saw a winter landscape with snow (see Figure 8.1). When asked to pick a card with an image to match, the right cortex, which controls the left hand, picked a shovel to go with the theme of snow, while the left cortex, which controls the right hand, picked an image of a chicken to go with the claw. This shows that each side of the cortex could recognize its image and make an appropriate association. When the patient was asked why he chose those two images, the response came from the left side (that is the only one that can speak—the right is mute); the response was, "Oh, that's

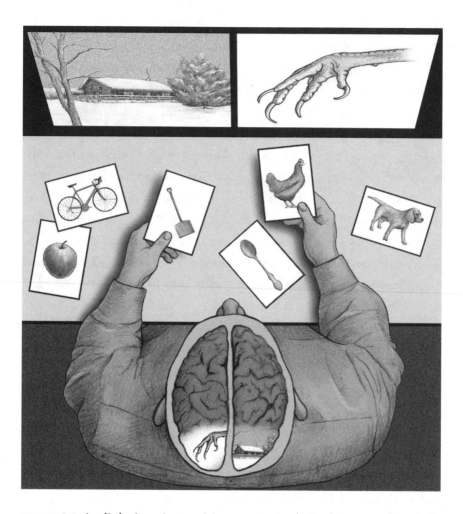

FIGURE 8.1. A split-brain patient receiving separate visual stimulation to each cortical hemisphere. The chicken claw in the right visual field activates the left hemisphere while the shed in the left visual field activates the right hemisphere. When the patient is asked to explain his choices of thematically related images, confabulation results. *Joan M. K. Tycko, illustrator.*

simple. The chicken claw goes with the chicken and you need a shovel to clean out the chicken shed."

Let's think carefully about what's happening here. The left brain saw the chicken claw but not the snow scene. When faced with the shovel and the chicken it retroactively constructed a story to make these disparate choices appear to make sense. Michael Gazzaniga, in his book entitled *The Mind's Past*, from which this example is taken, notes, "What is amazing here is that the left hemisphere is perfectly capable of saying something like, "Look, I have no idea why I picked the shovel . . . Quit asking me this stupid question." Yet it doesn't.

Here's another brief example, also from Gazzaniga. If, in a split-brain patient, the (mute) right brain receives the instruction "Go take a walk," the subject will push the chair back and prepare to leave. If, at that point, the (speaking) left brain, which had no access to the instruction, is asked "What are you doing?" it will manufacture a seemingly coherent response to make sense of the body's action, such as "I was feeling thirsty and decided to go get a drink" or "I had a cramp in my leg and needed to work it out." This is not just a fluke of one or two split-brain patients. The narrative-constructing capacity of the left cortex has now been clearly observed in more than 100 split-brain patients in many different situations.

In all humans, not just those who have had split-brain operations, this action of the left cortex is revealed in narrative dreaming. Why do we have *narrative* dreams at all? If the underlying purpose of these dreams is indeed memory consolidation and integration, then why don't we just experience isolated vignettes or flashes of memory instead of an unfolding bizarre and illogical story? The answer is that the narrative-constructing function of the left cortex cannot be switched off, even during sleep. Like the cerebellar system designed to reduce perception of self-generated movements discussed in Chapter 1, it is always on whether or not its function is relevant for the particular task at hand. The dream researcher David Foulkes tells Andrea Rock (in her book *The Mind at Night*, p. 127), "The interpreter [the narrative-constructing function of the left cortex] is doing an even more spectacular job of story-making than it does in waking, because the brain in sleep is activated but the raw material it has to work with is much different. You lose yourself, you lose the world and thought is no longer directed."

I SUGGEST THAT the left cortex's always-on narrative-constructing function promotes the acquisition of religious thought through both subconscious and conscious means. Religious ideas largely involve nonnaturalistic explanations. Whether religious ideas are regarded by their practitioners as "faith" or merely "given knowledge," they share the property that they violate everyday perceptual and cognitive structures and categories. The left cortex predisposes us to create narratives from fragments of perception and memory. Religious ideas are similarly formed by transforming everyday perceptions, by building coherent narratives that bridge otherwise disparate concepts and entities. Pascal Boyer proposes that the most effective religious concepts preserve all the relevant inferences of a particular cognitive category except those that are specifically prohibited by a special nonnaturalistic aspect.

> There has only been one woman in history who has given birth without having sex and we worship her for that reason.
>
> Category: person. Special aspect: virgin birth.
>
> There is one god who is all-powerful and all-knowing and who can hear your thoughts. You can pray to our god in a temple or anywhere else.
>
> Category: person. Special aspect: all-powerful, all-knowing.

Some ebony trees can recall conversations people hold in their shade. These can be revealed by burning a stick from the tree and interpreting the pattern of fallen ash.

Category: plant. Special aspect: recalls conversations.

The binding together of disparate percepts and ideas to create coherent narrative that violates our everyday waking experience and cognitive categories is a left cortical function that underlies both dreaming and the creation and social propagation of religious thought. This function operates subconsciously. We are not aware of the stories spun by our left cortex in our waking lives. We pay no attention to the man behind the curtain.

But in our narrative dreams, we have the experience of extended violations of conventional logic, connection, and causation. Our dreams give us nonnaturalistic experiences. They allow us to conceptualize systems and stories that are not constrained by our conventional waking categories and causal expectations. In this way, the subconscious process of narrative creation is made evident to our conscious mind.

So, I hypothesize that left cortex narrative creation works in two ways to promote religious thought: subconsciously, to make the cognitive leaps that underlie nonnaturalistic thought (that which violates categories, expectations, and causality) and consciously, through recollected dreams, to create a template for nonnaturalistic thought. In this vein, it is not an accident that cross-cultural ritual practice often incorporates dreaming, hallucinogenic drugs, trance, dance, meditation, and music. All of these aspects of ritual practice, by moving us away from waking consciousness, provide nonnaturalistic, dreamlike experience guided by the left cortex, and thereby reinforce the religious impulse.

Let's be clear about what I'm proposing here. Although all cultures have religious thought, ultimately, religious thought is an individual phenomenon. Within a given culture, individuals vary considerably in their religious thought with some claiming to have none whatsoever. I am not proposing that some individuals (or some cultural groups) have biological variation, either genetic or epigenetic, that predicts their predisposition for religious thought. Rather, I am claiming that, though on the individual level religion is a matter of personal faith as influenced by sociocultural factors, our shared human evolutionary heritage, as reflected in the structure and function of our brains, predisposes us as a species for religious thought in much the same way that it predisposes us for other human cultural universals such as long-term pair bonding, language, and music.

FAITH IS NOT only the province of religion. When John Brockman posed the question, "What do you believe is true even though you cannot prove it?" to a group of scientists and academics through his edge.org website, the answers were voluminous and wide-ranging, although mostly not explicitly religious. Even the most hard-headed rational atheists had an answer. On some level, all of us are hard-wired, or are at least strongly predisposed, to believe things we cannot prove. That essential act of faith is central to human mental function. It is an important initial step in making sense of our world.

So why, particularly in the United States, do scientific and religious ideas often lead to cultural war? One reason is that many scientists have not been very humble about this matter. Scientific investigation *has* strongly challenged the factual basis of some particular religious ideas (such as a Judeo-Christian biblical flood or a 6,000-year-old Earth or the creation of Eve from Adam's rib). For some scientists, such findings are enough to warrant a wholesale repudiation of religious faith and the faithful. But, scientifically speaking, is that warranted? Although the details of particular religious texts are falsifiable, the core tenets of many religions (a belief in a God, the existence of an immortal soul) are not. Science *cannot prove or disprove* the central ideas underlying most religious thought. When scientists claim to invalidate these core tenets of religious faith without the evidence to do so, they do a disservice to both science and religion.

On the religious side there has often been a similar intolerance of scientific thought. Fundamentalists of many religions (Christians, Muslims, Jews, and Hindus to name but a few) have insisted upon a rigid, literal reading of sacred texts. For these people, the rejection of science is a given, and the firmer the rejection the better, because strong rejection is seen as an opportunity to demonstrate the strength of their religious feeling: "I have faith, I believe this in my heart and nothing you can say or do can shake my faith." Attempts to reconcile literal readings of sacred texts with observations of the world often lead to improbable proposals. For example, an exhibit at the Creation Museum in Petersburg, Kentucky, shows pairs of dinosaurs marching up the ramp to board Noah's Ark!

Of course, most fundamentalists don't get really worked up about most areas of science. Chemistry and mathematics remain largely unmolested. Physics does not inspire impassioned debates at school board meetings (although this may change). Evolutionary biology takes most of the heat. It's not just that evolutionary biology contradicts traditional creation stories such as that in the Book of Genesis. There has also been an assumption that if one accepts the idea that life developed without divine intervention, it necessarily follows that all aspects of religious thought must be rejected. Those who take this line of argument to extremes argue that when religious thought is rejected moral and social codes will degenerate, and "the law of the jungle" will be all that is left. It is imagined by religious fundamentalists that those who do not share their particular religious faith are incapable of leading moral lives.

The tragedy is that these suppositions are simply not true. One *can* be a person of faith and still accept a scientific worldview, including evolutionary biology (one can also be a scrupulously moral agnostic or atheist). It's only *fundamentalist* religion that is incompatible with science. Fortunately, many of the world's religious leaders have accepted the idea that scientific and religious ideas need not be mutually exclusive. His Holiness the Dalai Lama has said, "If science proves Buddhism is wrong, then Buddhism must change." In stark contrast to fundamentalist Christian teachings, the Catholic bishops of England, Scotland, and Wales have stated, "We should not expect to find in Scripture full scientific accuracy or complete historical precision." They hold that the Bible is true in passages relating to human salvation and the divine origin of the soul but, "we should not expect total accuracy from the Bible in other, secular matters" (from *The Gift of Scripture,* published by the Catholic Truth Society, United Kingdom). The Vatican has essentially stated that the scientific consensus model of evolution is valid but that it explains only the biological part of humanity, not the spiritual mystery. How utterly reasonable!

WE ALL BELIEVE some things we cannot prove. Those unproven ideas that are ultimately subject to falsifying experiment or observation constitute "scientific faith." Those that are not constitute religious faith. These two modes of thought are not mutually exclusive, as fundamentalist religious leaders and some scientists would have you believe. Rather, they are two branches of the same cognitive stream. Our brains have evolved to make us believers.

Chapter Nine

The Unintelligent Design of the Brain

HOSTILITY TO EVOLUTIONARY biology has been a feature of certain parts of the American political and religious landscape for many years, although it has been much less of an issue in most other countries. Most religious denominations and indeed most Christian leaders have made their peace with the basic tenets of evolution: that all present life on Earth derives from a common 3.5-billion-year-old ancestor, and that living things change slowly through a random process of genetic mutation coupled with natural selection. Indeed, Pope John Paul II made this point in a 1996 address to the Pontifical Academy of Sciences entitled "Truth Cannot Contradict the Truth." He said, "Today, almost half a century after the publication of the encyclical [a previous statement from Pope Pius XII in 1950 that said there was no opposition between evolution and the doctrine of the faith], new knowledge has led to the recognition of the theory of evolution as more than a hypothesis."

But fundamentalist Christians adhere to a literal reading of the Book of Genesis and have for many years sought to have this biblical view taught in American public schools. When these attempts were repeatedly banned by the courts on the basis of the Constitutional separation of church and state, a new strategy was born called "scientific creationism." A group of fundamentalist American Christians attempted to claim that careful examination of the geological and biological record supports the story of Genesis—that the Earth is 6,000 years old, that all species were created simultaneously, and that mass extinctions seen in the fossil record were caused by the Noah's flood. But this argument also failed. Not only was it impossible to marshal the evidence to support these claims scientifically, but, in the words of the evolutionary biologist Jerry Coyne, "American courts clearly spied clerical collars beneath the lab coats" and struck down teaching of so-called scientific creationism in schools.

In the 1990s yet another strategy was developed. Recognizing that explicit references to religion would always be rejected by the courts, a group of fundamentalist Christian academics took a step back and sought to devise a theory that would challenge evolutionary biology but would appear to be scientifically reasonable. This movement, dubbed "intelligent design," does not try to provide support for such obviously scientifically untenable points as a 6,000-year-old Earth, Noah's flood, or other aspects of the Genesis story. In fact, when talking to the world at large, the supporters of intelligent design are careful not to mention God or religion at all. Rather, they claim that living creatures are just too intricate and clever to have arisen by random mutation and selection. These forms, they say, are too elegant and too complex to attribute to anything other than a very clever designer. Therefore, an unspecified intelligent designer must be at work. In this way of thinking, gradual change of living things is admitted and the fossil record and the genetic relationships between living organisms can be accounted for, but the engine driving this change is challenged.

The crux of the matter is this: intelligent design purports to be a scientific theory, but it isn't. Pope John Paul II hit one out of the ballpark when he offered the following definition. "A theory is a metascientific elaboration distinct from the results of observation but consistent with them. By means of it, a series of independent data and facts can be related and interpreted in a unified explanation. A theory's validity depends on whether or not it can be falsified. It is continually tested against the facts; wherever it can no longer explain the latter, it shows its limitations and unsuitability. It must then be rethought" (address to the Pontifical Academy of Sciences, October 23, 1996).

Evolution is a theory. It can be falsified by particular findings, such as a hominid skeleton dated to the Jurassic Era. Intelligent design is not. It rests on a subjective inference of design that is not subject to a falsifying experiment or observation. It is not surprising that despite lavish funding from certain religious and political groups, the intelligent design movement has provided no fieldwork or laboratory experimentation to bolster its claims. Yes, books are written, papers are presented and published, and even mathematical models are constructed. All the trappings of science are there, but there is no science at the core.

IS THE GOAL of the intelligent design movement really to do legitimate science to challenge the theory of evolution, or is its goal merely to craft a sufficiently watered-down view of creationism to appear scientific and thereby gain a place at the debating table and fly under the radar of the courts? Although intelligent design proponents are careful not to mention religion in public hearings or debates, quite a different picture emerges when they are addressing fundamentalist Christian audiences. Phillip E. Johnson of the University of California at

Berkeley, one of the founders of the intelligent design movement, said, "Our strategy has been to change the subject a bit so that we can get the issue of intelligent design, which really means the reality of God, before the academic world and into the schools" (American Family Radio, January 10, 2003). William Dembski of the Southern Baptist Theological Seminary, another well-known intelligent design proponent, has stated, "Intelligent design readily embraces the sacramental nature of physical reality. Indeed, intelligent design is just the Logos theology of John's Gospel restated in the idiom of information theory" (*Touchstone: A Journal of Mere Christianity*, July 1999).

In its public face, intelligent design has been cleverly crafted to appear as a legitimate scientific theory with no ties to a specific religious agenda. This gives political cover to politicians and school board members who can adopt a tone of fairness in saying, "Let's present our students with both sides of this fascinating scientific debate." For example, in March 2002, U.S. Senator Rick Santorum (Republican of Pennsylvania) said, "Proponents of intelligent design are not trying to teach religion via science, but are trying to establish the validity of their theory as a scientific alternative to Darwinism." In August 2005, President George W. Bush weighed in: "Both sides ought to be properly taught . . . so people can understand what the debate is about."

IF YOU BELIEVE that life was designed by an intelligent force (whether that be the Judeo-Christian God, angels, the Allah of Islam, or even extraterrestrials), then the human brain, the presumed seat of reason, morality, and faith, is the obvious test case to reveal this design. After all, this 2.5-pound lump of tissue can solve problems in recognition, categorization, social interaction, and many other areas that routinely baffle the world's most sophisticated supercomputers. These supercomputers are designed and programmed by teams of extremely talented hardware and software engineers. Doesn't this imply that the brain was designed by an even more skillful engineer?

The proponents of intelligent design make two main arguments. First, as we have seen, they contend that living things could not have arisen through Darwinian evolution because they contain structures that are "irreducibly complex." This means that if you remove any one component part from one of these structures (such as an ion channel or a bacterial flagellum), it will not be partially crippled, but rather will fail to function entirely. Therefore, how can we imagine that these structures arose from random change and gradual selection when the intermediate forms would fail? Second, they claim that random mutation and selection cannot generate new information and therefore cannot produce the "specified complexity" necessary to adapt to the environment. In their view, only an intelligent agent can get around these problems.

Specialists in molecular evolution and information theory (I am neither) have refuted these core claims in exquisite detail (see the Further Reading and Resources section of this book). To me the most compelling evidence against the claim of irreducible complexity is that careful observation reveals that complexity is not irreducible at all. For example, the flagellum (a whip-like structure that bacteria spin to move through liquids) of more recent bacteria is more complex than the flagellum of more ancient bacteria. In many cases, complex structures such as the flagellum arise when genes mediating other functions (such as ion pumps for example) are randomly duplicated in the genome, and then one copy of the duplicated gene accumulates mutations that allow it to adopt a new function (as a component of the flagellum).

The claim that "specified complexity" cannot arise through random mutation and selection is a specious argument. The critique from information theory, that new information cannot be generated by the evolutionary process, would hold only if the evolutionary process were charged with matching an independently given pattern. This is not the case. The evolutionary process does not strive to build prespecified complex structures such as eyes, kidneys, or brains. It has no goal. The only driving force of evolution is reproductive success and the related issues of kin selection and the reproductive success of one's offspring. If building complex structures increases reproductive fitness, then they may arise, but if destroying complex structures also increases reproductive fitness, then complex structures can just as easily be destroyed or altered (as happened when the eyes of cave-dwelling fish became nonfunctional).

So, where does this leave the intelligent design movement? Essentially, it leaves its proponents saying: *"Look* at that thing. It's just too *cool* not to have been actively designed."* Michael Behe, a biochemist, took this line when defending the intelligent design movement on the *New York Times* op-ed page (February 7, 2005): "The strong appearance of design allows a disarmingly simple argument: If it looks, walks and quacks like a duck, then, absent compelling evidence to the contrary, we have warrant to conclude it's a duck. Design should not be overlooked simply because it's so obvious." Behe would like intelligent design to be the default explanation for biological structure, with the burden placed on any competing explanation to prove otherwise (Figure 9.1).

Is the evidence for design in biological systems so obvious? I hold that the brain, the ultimate test case, is, in many respects, a true design nightmare. Let's review a bit. When we compare the human brain to that of other vertebrates, it becomes clear that the human brain has mostly developed through agglomeration. The difference between the lizard brain and the mouse brain does not involve a wholesale redesign. Rather, the mouse brain is basically the lizard brain with some extra stuff thrown on top. Likewise, the human brain is basically the mouse brain with still more stuff piled on top. That's how we wind up with two visual systems and two auditory systems (one ancient and one mod-

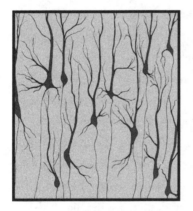
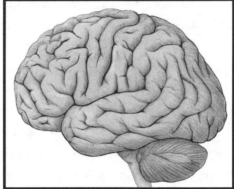

FIGURE 9.1. Do these all reveal the work of an intelligent designer? Proponents of intelligent design are fond of using the carvings of presidents on Mount Rushmore (top left) as an example of something that we can tell was intelligently designed even without any specific experiments. They would say that evidence of intelligent design is equally evident in biological structures such as the shell of the chambered nautilus (shown in cross section in the top right), neurons in the cerebral cortex (bottom left), and the brain as a whole (bottom right). *Joan M. K. Tycko, illustrator.*

ern) jammed into our heads. The brain is built like an ice cream cone with new scoops piled on at each stage of our lineage.

Accidental design is even more obvious at the cellular level in the brain. The job of neurons is to integrate and propagate electrical signals. Yet, in almost all respects, neurons do a bad job. They propagate their signals slowly (a million times more slowly than copper wires), their signaling range is tiny (0 to 1,200 spikes/second), they leak signals to their neighbors, and, on average, they successfully propagate their signals to their targets only about 30 percent of the time. As electrical devices, the neurons of the brain are extremely inefficient.

So, at either the systems or the cellular level, the human brain, which the intelligent design crowd would imagine to be the most highly designed bit of tissue on the planet, is essentially a Rube Goldberg contraption. Not surprisingly, some proponents of intelligent design have left themselves a way to retreat on this point. Michael Behe writes, "Features that strike us as odd in a design might have been placed there by the designer for a reason—for artistic reasons, for variety, to show off, for some as-yet-undetectable practical purpose or for some unguessable reason—or they might not." Or, stated another way, if on first glimpse biological systems look cool, that must be the result of intelligent design. If, on closer inspection, biological systems look like a cobbled-together contraption, that's still got to be from intelligent design, just intelligent design with an offbeat sense of humor. Clearly, this position is not a true, falsifiable scientific hypothesis, as is the theory of evolution. The idea of intelligent design is merely an assertion.

WHAT HAPPENS WHEN we lift the hood, so to speak? After all, the complete sequences of the human, mouse, worm, and fly genomes are all in hand. What can they tell us? They make the case for evolution much stronger. Would you like to see genes duplicated to underlie development of novel complex traits?

They're there. How about disabled genes which have mutated to the point where they no longer function to make protein (called pseudogenes)? Check. Genes where mutations have accrued across species to give rise to new functions? No problem.

The complete sequences of these genomes have been available only for a few years and there's a lot we don't know about how genes direct the structure and function of tissues and how their expression responds to environmental cues. Our knowledge of gene-environment interactions in forming the structure and function of the brain is at a very early stage. Nonetheless, there are some striking examples of how variation in gene structure underlies brain structure. One of the best is the ASPM gene (previously mentioned in Chapter 3). Recall that this gene, which codes for a protein in the mitotic spindle (a structure used to organize the chromosomes during cell division), seems to determine how many times cortical progenitor cells divide before they become committed to becoming cortical neurons. As a result, this gene is crucial for determining cortical size. Humans who harbor certain mutations in ASPM will be microcephalic. You may also recall that an important part of this protein is a segment that binds the messenger molecule calmodulin and that the calmodulin-binding region is present in two copies in the ASPM gene of the roundworm, 24 in the fruit fly, and 74 in humans. Analysis of the ASPM gene in chimpanzees, gorillas, orangutans, and macaque monkeys has indicated that change in this gene, particularly in its calmodulin-binding region, has been particularly accelerated in the great ape family. These findings strongly indicate that the ASPM gene is one major determinant in the evolution of cortical size. In a few more years we will not have to speculate about the genes underlying the evolution of brain structure. We will have them in hand.

So, with genomic information supporting evolution, including brain evolution, so strongly, where are the intelligent design advocates to retreat on this

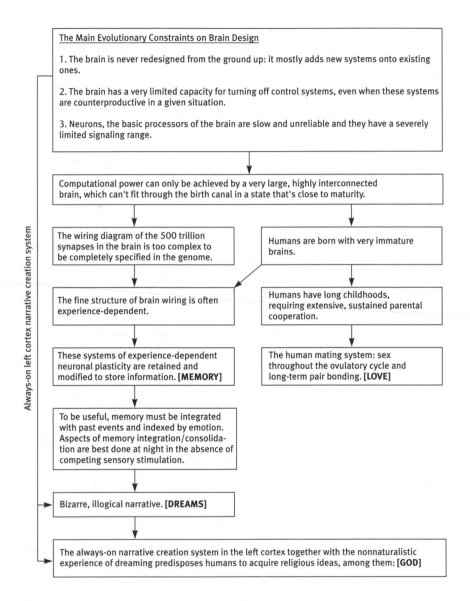

Always-on left cortex narrative creation system

The Main Evolutionary Constraints on Brain Design

1. The brain is never redesigned from the ground up: it mostly adds new systems onto existing ones.

2. The brain has a very limited capacity for turning off control systems, even when these systems are counterproductive in a given situation.

3. Neurons, the basic processors of the brain are slow and unreliable and they have a severely limited signaling range.

Computational power can only be achieved by a very large, highly interconnected brain, which can't fit through the birth canal in a state that's close to maturity.

The wiring diagram of the 500 trillion synapses in the brain is too complex to be completely specified in the genome.

Humans are born with very immature brains.

The fine structure of brain wiring is often experience-dependent.

Humans have long childhoods, requiring extensive, sustained parental cooperation.

These systems of experience-dependent neuronal plasticity are retained and modified to store information. **[MEMORY]**

The human mating system: sex throughout the ovulatory cycle and long-term pair bonding. **[LOVE]**

To be useful, memory must be integrated with past events and indexed by emotion. Aspects of memory integration/consolidation are best done at night in the absence of competing sensory stimulation.

Bizarre, illogical narrative. **[DREAMS]**

The always-on narrative creation system in the left cortex together with the nonnaturalistic experience of dreaming predisposes humans to acquire religious ideas, among them: **[GOD]**

FIGURE 9.2. Love and memory, dreams and God: a chart encapsulating the main argument of the book.

point? Behe points to a way out: Imagine that an intelligent designer assembled some simple organisms long ago, and then washed her hands of the whole thing and let evolution take over. In this way you may still have evolved from a common ancestor with chimpanzees and mice and flies and worms, but this was only allowed by an intelligent design that ended more that 600 million years ago. By positing that both creation by intelligent design and evolution have occurred, and that the designer's intent is unfathomable owing to an offbeat sense of humor, Behe has carved out a tiny mountaintop niche. This is a scrap of rhetorical ground that is impossible to attack, but from which nothing can be launched, either. Needless to say, it's not falsifiable and therefore it's not a genuine scientific theory. Not surprisingly, many others in the intelligent design movement (William Dembski and Phillip Johnson among them) are not willing to give up so much ground. They hold onto the notion that Darwinian evolution is incapable of building anything useful.

Perhaps the problem is the gee-whiz factor. It is indeed deeply and profoundly amazing that there is a tissue such as the human brain that confers our very humanness. It's not surprising that, for some, pondering the awe-inspiring concept of the mind-in-the-brain leads to a religious, faith-based (untestable, unfalsifiable) explanation rather than a scientific, faith-based (testable, falsifiable) hypothesis. What's interesting here is that though there are many different ways to get the story wrong, the intelligent design group has gotten it exactly, explicitly 180 degrees wrong. The transcendent aspects of our human experience, the things that touch our emotional and cognitive core, were not given to us by a Great Engineer. These are not the latest design features of an impeccably crafted brain. Rather, at every turn, brain design has been a kludge, a workaround, a jumble, a pastiche. The things we hold highest in our human experience (love, memory, dreams, and a predisposition for religious thought; Figure 9.2) result from a particular agglomeration of ad hoc solutions that have been piled on through millions of years of evolutionary history. It's not that we have fundamentally human thoughts and feelings *despite* the kludgy design of the brain as molded by the twists and turns of evolutionary history. Rather, we have them precisely *because* of that history.

That Middle Thing

THERE ARE A lot of fascinating topics in brain function that I didn't cover here, including language, brain aging and disease, psychoactive drugs, hypnosis, and the placebo effect. This is all great stuff and it took a lot of topical self-control to prevent this book from becoming a huge tome. More important, I think that you have discerned by now that many of the explanations present-day biology can offer about higher brain function are rather incomplete. But there are a few lovely examples where an explanation in terms of molecules and cells provides a nearly complete understanding of something we experience. One of my favorites is the finding that people equate the sensation of chili peppers (either in the mouth or on the skin) with the sensation of heat. At first, you might imagine that this is merely an example of metaphorical speech that has arisen in a few cultures. Not so: in every culture where people are exposed to capsaicin, the active ingredient in chili peppers, they characterize the sensation as "hot," suggesting a biological basis. OK then, you might say, perhaps the answer is that in your palate there are temperature-sensing neurons and also some capsaicin-sensing neurons and these two types of neuron project to the same place in the brain, which, when activated, gives a sensation of "heat." It turns out that this explanation's not quite right either. The real story is that there is a family of receptors for capsaicin and related compounds that are located on nerve endings in the mouth (and other places such as the skin). These are called vanilloid receptors (vanilloids are the class of chemical that include capsaicin and related compounds), and they are activated by both capsaicin and warming, giving rise to similar sensations of "heat" for both stimuli. In this case, the experience at the level of behavior is almost entirely explained at the level of the single receptor molecule. This even reveals why if you drink hot tea after a spicy meal the tea seems extra-hot: the receptor is super-activated by warming and capsaicin together! And yes, it's not just chili pepper heat that can be explained in this way: an analogous story concerns a receptor family called the cold/menthol receptors, which give rise to the cross-cultural association between coolness and the active ingredient of mint.

Unfortunately, most of the biological explanations for experience and behavior are not nearly this neat and compact. For example, in Chapter 5, I talked about learning and memory. We know that if you get a hole in your hippocampus you can't store new memories for facts and events. We also know this seems to require a chemical process at synapses in your hippocampus whereby certain patterns of neural activity result in activation of NMDA-type glutamate receptors. This receptor activation, in turn, sets in motion a series of chemical steps that render these activated synapses weaker or stronger and keeps them that way for a long time—a phenomenon called long-term synaptic depression and potentiation, LTP and LTD. And we know this molecular phenomenon seems to underlie declarative memory, because injection into the hippocampus of drugs that prevent the activation of NMDA-type glutamate receptors makes it impossible for animals to store new memories for facts and events.

At first glance, this would appear to be a fairly complete explanation, but it's not. What's missing is *that middle thing*. How is it that changing the strength of some synapses in the hippocampal circuit actually gives rise to memories for facts and events, as recalled during behavior? We have a molecular explanation for how synapses get weaker and stronger. At a behavioral level, we can show that interfering with this molecular process (and perhaps some other things as well) disrupts memory. But our understanding of the intervening steps is almost nonexistent: that middle thing is a big, nasty, embarrassing gap for brain scientists. Unfortunately, the middle-thing problem is not confined just to learning and memory. Similar gaps between molecules and behavior exist for our understanding of many other complex cognitive and perceptual phenomena.

I DON'T WANT to end on a downer. Brain science has made huge progress in identifying the molecular and cellular underpinnings of behavior and experience. In most cases, a complete gap-free explanation that flows from molecules to behavior with an intermediate-level understanding of systems and circuits is not yet in hand. But let's talk about an example where it seems possible to find the brain scientist's holy grail—a gap-free explanation of a high-level process in the brain. In this case, the process is a particular form of learning that involves control of your eye muscles.

Here I'm talking about a learning task that's similar to the one given to Pavlov's famous dog. You'll recall that Pavlov's dog had no particular response to the sound of a bell and would reflexively salivate when presented with meat. After many experiences when the bell was rung immediately before giving meat, the dog learned to associate these two stimuli such that the sound of the bell alone would cause the dog to salivate. Psychologists have named this simple form of learning classical conditioning. It is a type of nondeclarative memory. Now if you (or a rat, rabbit, or a mouse) are brought into the lab and you hear a bell (or some other innocuous sound) at moderate volume, you will have no particular behavioral response. If a brief puff of air is delivered to your eye, you will blink reflexively. You don't think about it; it will just happen, as when the doctor taps your knee with the hammer to make your leg jump during a physical exam. If, however, the bell tone is presented for a half-second or so and at the end of this tone a puff of air is directed to your eye, you will begin to learn to associate the bell tone with the air puff just as Pavlov's dog did for the meat and bell. What this means is that after many pairings of air puff and bell tone, you will blink your eye in response to the tone alone so that your eyelid is closed when the air puff would be expected to arrive. This form of learning, which is called associative eyelid conditioning, absolutely requires that the tone predicts the arrival of the air puff. If you experience tones alone or air puffs alone or even both air puffs and tones but delivered out of synch, you will not learn. After you learn this response, it is completely subconscious and beyond your control— you can't help blinking when you hear the tone.

Many labs all over the world have been working for a long time to understand how this simple form of learning happens, and a lot of progress has been made. We know, for example, that the air puff activates a small group of neurons in a portion of the brain called the inferior olive (yes, that's really what it's called—the early anatomists who named this stuff were given to flights of fancy). If, in a rabbit, you artificially activate this location in the brain with an electrode, it can replace the air puff during training. The bell tone, on the other hand, activates a group of cells in the brainstem that give rise to a set of axons called mossy fibers. In a way similar to what was found for the air puff, you can replace the bell tone during training with artificial electrical activation of these mossy fibers. So in order to store the memory for associative eyelid conditioning, the bell tone signals and the air puff signals must meet somewhere in the brain, and when they arrive together (but not separately) they must produce a change in the neural circuit that ultimately causes a blink in response to the bell tone.

Figure E.1 shows how this might occur. The tone and air puff signals are both received in the cerebellum (that baseball-sized blob hanging off the back of your brain that is important for motor coordination). In particular, these signals both excite a fan-shaped class of neuron called the cerebellar Purkinje cell. The air puff signal comes directly through climbing fibers, but the bell tone signal comes indirectly: the mossy fibers excite cerebellar granule cells, and the axons of the granule cells, called parallel fibers, in turn, excite Purkinje cells. When climbing fiber and parallel fibers are activated together, as occurs when the bell tone and air puff are paired, and this is repeated many times, the result is a long-lasting decrease in the strength of those excitatory parallel fiber–Purkinje cell synapses activated by bell tones; this is called cerebellar long-term synaptic depression, or cerebellar LTD.

It turns out that we now know a lot about the molecular alterations that underlie cerebellar LTD. The synapses are made weaker by changes on the postsynaptic side that result in triggering the internalization of neurotransmitter receptors, thereby rendering them unavailable to bind the neurotransmitter (which in this case is glutamate) at the cell surface. We understand some details of this process in excruciating molecular detail. For example, the main form of the glutamate receptor at this synapse is composed of a chain of 883 amino acids, and the crucial molecular step in triggering internalization of the receptor is the transfer of a phosphate group from an enzyme called protein kinase C to the number 880 amino acid, which happens to be a serine.

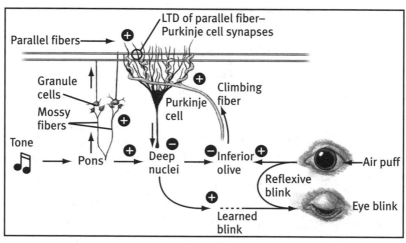

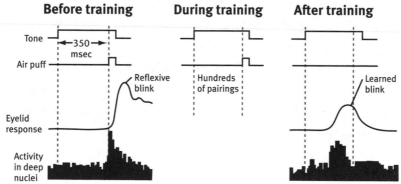

FIGURE E.1. A proposed circuit-level explanation for a simple form of learning called associative eyelid conditioning. See the text for more detail. After repeated pairing of a tone and an air puff, the animal learns that the tone is predictive of the air puff, and it will reflexively blink in response to the tone alone. Tone–air puff pairing is thought to produce long-term depression (LTD) of the excitatory parallel fiber–Purkinje cell synapse. This ultimately results in an increase in tone-driven activity in the deep nuclei, and it is this activity that drives the learned blink. Adapted from D. J. Linden, From molecules to memory in the cerebellum, *Science* 301:1682–1685 (2003). *Joan M. K. Tycko, illustrator.*

So how do we get from cerebellar LTD to a learned blink in response to tone? When the parallel fiber synapse is depressed as a result of tone–air puff pairing, it produces less excitation of the Purkinje cell. The Purkinje cell, in turn, fires less. Because the Purkinje cell is inhibitory, the cells that receive contacts from its axon are less inhibited and therefore fire more in response to tones. This occurs in a particular place called the cerebellar interposed nucleus, where recordings of neural activity have shown that as rabbits learn the tone–air puff association, firing rates in the interval between the start of the tone and the start of the air puff gradually increase. What's more, artificial stimulation of the appropriate portion of the interposed nucleus can itself give rise to eye blinks.

Now, this is a model and, with further experiments, portions of it will be proved incomplete or even wrong. But the exciting thing about this explanation is that there's no missing middle thing. Here is one of the very few examples in the brain where it is possible to go from a detailed molecular description of a change at a synapse, through an anatomically well-defined wiring diagram, to a high-level behavior, in this case, a form of nondeclarative memory. The payback comes from being willing to study a behavior that's simple (memory for rules and procedures) as opposed to one where it is still too difficult to grasp the middle thing (such as the problem of memory for facts and events).

The holy grail of complete biological explanations for behavior is not in hand, but it is emerging for some simple phenomena. We neurobiologists are an optimistic lot by nature, but there is every reason to believe that our level of understanding will continue to increase rapidly. Furthermore, it is very likely that working out complete molecules-to-circuits-to-behavior explanations for some simple forms of learning such as associative eyelid conditioning will yield some general principles and insights that can then be applied to more complex phenomena.

So, the next time you hear some misguided congressman give a spittle-strewn rant about "how those ivory-tower pointy-headed scientists are spending our tax dollars to figure out how a rabbit learns to blink," you can fire off an e-mail explaining exactly why this line of work is crucial for understanding the molecular basis of cognition and diseases of memory: it's a step in conquering the next great scientific frontier.

Further Reading and Resources

Acknowledgments

Index

Further Reading and Resources

1. The Inelegant Design of the Brain

MATERIAL FOR A GENERAL AUDIENCE

Carter, R. 1998. *Mapping the Mind.* University of California, Press, Berkeley. For my money, this is the best coffee-table book written about brain function to date. It's clear, scientifically accurate, and has lovely illustrations.

Ramachandran, V. S., and Blakeslee, S. 1998. *Phantoms in the Brain.* William Morrow, New York. This is my favorite of the "illuminate higher brain functions through interesting neurological case studies" genre. It does a good job of blending the case studies with human laboratory experiments and a dash of philosophy and intellectual history.

SCIENTIFIC REPORTS AND REVIEWS

Blakemore, S. J., Wolpert, D., and Frith, C. 2000. Why can't you tickle yourself? *Neuro-Report* 11:11–16.

Corkin, S. 2002. What's new with the amnesic patient H.M.? *Nature Reviews Neuroscience* 3:153–160.

Shergill, S. S., Bays, P. M., Frith, C. D., and Wolpert, D. M. 2003. Two eyes for an eye: the neuroscience of force escalation. *Science* 301:187.

Weiskrantz, L. 2004. Roots of blindsight. *Progress in Brain Research* 144:229–241.

2. Building a Brain with Yesterday's Parts

MATERIAL FOR A GENERAL AUDIENCE

Nicholls, J. G., Wallace, B. G., Fuchs, P. A., and Martin, A. R. 2001. *From Neuron to Brain*, 4th ed. Sinauer, Sunderland, MA. There isn't much on molecular and cellular neurobiology for a general audience. This, in my view, is the best college textbook on the subject.

3. Some Assembly Required

MATERIAL FOR A GENERAL AUDIENCE

Ridley, M. 2003. *Nature via Nurture.* Harper Perennial, New York. A splendid, very well written book on the nature-versus-nurture debate in human brain development. The author takes the reasonable middle path. A page turner, actually.

SCIENTIFIC REPORTS AND REVIEWS

Bouchard, T. J., Jr., and Loehlin, J. C. 2001. Genes, evolution, and personality. *Behavioral Genetics* 31:243–273.

Bradbury, J. 2005. Molecular insights into human brain evolution. *PLoS Biology* 3:E5. *PLoS Biology* is an open-access journal whose contents are available free to all at www.plos.org.

Kouprina, N., Pavlicek, A., Mochida, G. H., Solomon, G., Gersch, W., Yoon, Y. H., Collura, R., Ruvolo, M., Barrett, J. C., Woods, C. G., Walsh, C. A., Jurka, J., and Larionov, V. 2004. Accelerated evolution of the ASPM gene controlling brain size begins prior to human brain expansion. *PLoS Biology* 2:E126.

Meyer, R. L. 1988. Roger Sperry and his chemoaffinity hypothesis. *Neuropsychologia* 36:957–980.

Verhage, M., Maia, A. S., Plomp, J. J., Brussaard, A. B., Heeroma, J. H., Vermeer, H., Toonen, R. F., Hammer, R. E., van den Berg, T. K., Missler, M., Geuze, H. J., and Südhof, T. C. 2000. Synaptic assembly of the brain in the absence of neurotransmitter secretion. *Science* 287:864–869.

4. Sensation and Emotion

MATERIAL FOR A GENERAL AUDIENCE

Ramachandran, V. S., and Hubbard, E. M. 2003. Hearing colors, tasting shapes. *Scientific American* 288:52–59.

Stafford, T., and Webb, M. 2004. *Mind Hacks.* O'Reilly, Sebastopol, CA. This improbable, wonderful book from the computer book publisher O'Reilly is in a series of books with titles such as *Google Hacks* and *Linux Hacks.* Although it is a bit strange to shoehorn a brain book into this "computer tips and tricks" format, the end result is a fascinating collection of exercises you can do at home that reveal aspects of brain organization. It's particularly strong on sensory systems and is richly endowed with links to websites that support the "hacks" with Java applets, flash animation, and so on.

www.michaelbach.de/ot/index.html. This website shows 53 optical illusions, many of them animated. There is good commentary about the neural phenomena thought to underlie these illusions and references to the original scientific papers written about them.

www.prosopagnosia.com. This website about face-blindness is written by a woman named Cecilia Burman, who has this condition. It is particularly interesting for her descriptions of living with prosopagnosia and the strategies she uses to adapt in social situations.

SCIENTIFIC REPORTS AND REVIEWS

Beeli, G., Esslen, M., and Jancke, L. 2005. Synaesthesia: when coloured sounds taste sweet. *Nature* 434:38.

Eisenberger, N. I., and Lieberman, M. D. 2004. Why rejection hurts: a common neural alarm system for physical and social pain. *Trends in Cognitive Science* 8:294–300.

Nunn, J. A., Gregory, L. J., Brammer, M., Williams, S. C., Parslow, D. M., Morgan, M. J., Morris, R. G., Bullmore, E. T., Baron-Cohen, S., and Gray, J. A. 2002. Functional magnetic resonance imaging of synesthesia: activation of V4/V8 by spoken words. *Nature Neuroscience* 5:371–375.

Ramachandran, V. S. 1996. What neurological syndromes can tell us about human nature: some lessons from phantom limbs, Capgras syndrome, and anosognosia. *Cold Spring Harbor Symposium in Quantitative Biology* 61:115–134.

Ramachandran, V. S., and Hubbard, E. M. 2001. Psychophysical investigations into the neural basis of synaesthesia. *Proceedings of the Royal Society: Biological Sciences* 268:979–983.

Rizzolatti, G., and Craighero, L. 2004. The mirror-neuron system. *Annual Review of Neuroscience* 27:169–192.

Thilo, K. V., and Walsh, V. 2002. Chronostasis. *Current Biology* 12:R580–581.

Villemure, C., and Bushnell, M. C. 2002. Cognitive modulation of pain: how do attention and emotion influence pain processing? *Pain* 95:195–199.

Yarrow, K., and Rothwell, J. C. 2003. Manual chronostasis: tactile perception precedes physical contact. *Current Biology* 13:1134–1139.

5. Learning, Memory, and Human Individuality

MATERIAL FOR A GENERAL AUDIENCE

Le Doux, J. 2002. *Synaptic Self.* Penguin, New York. A well-argued book about the current state of research on the cellular basis of memory. It is particularly strong in considering the role of the amygdala in fear memory, which is the author's specialty.

Schacter, D. L. 2001. *The Seven Sins of Memory.* Houghton Mifflin, Boston. A wonderful, lucid book that describes the ways in which memory fails in healthy humans. These are explanations at the level of behavior and brain imaging, not the level of molecules and cells.

Squire, L. R., and Kandel, E. R. 1999. *Memory: From Mind to Molecules.* Scientific American Library, New York. Although I may argue with some of the details in the cellular/molecular part of this book, there is no denying that is does a nice job of encompassing the sweep of modern memory research. Nicely illustrated in the Scientific American style.

SCIENTIFIC REPORTS AND REVIEWS

Holtmaat, A. J., Trachtenberg, J. T., Wilbrecht, L., Shepherd, G. M., Zhang, X., Knott, G. W., and Svoboda, K. 2005. Transient and persistent dendritic spines in the neocortex in vivo. *Neuron* 45:279–291.

Malenka, R. C., and Bear, M. F. 2004. LTP and LTD: an embarrassment of riches. *Neuron* 44:5–21.

Morris, R. G., Moser, E. I., Riedel, G., Martin, S. J., Sandin, J., Day, M., and O'Carroll, C. 2003. Elements of a neurobiological theory of the hippocampus: the role of activity-dependent synaptic plasticity in memory. *Philosopical Transactions of the Royal Society of London, Series B, Biological Science* 358:773–786.

Nakazawa, K., McHugh, T. J., Wilson, M. A. and Tonegawa, S. 2004. NMDA receptors, place cells, and hippocampal spatial memory. *Nature Reviews Neuroscience* 5:361–372.

O'Keefe, J., and Nadel, L. 1978. *The Hippocampus as a Cognitive Map.* Oxford University Press, Oxford.

Zhang, W., and Linden, D. J. 2003. The other side of the engram: experience-dependent changes in neuronal intrinsic excitability. *Nature Reviews Neuroscience* 4:885–900.

6. Love and Sex

MATERIAL FOR A GENERAL AUDIENCE

Diamond, J. 1998. *Why Is Sex Fun?* Basic Books, New York. An excellent overview of human sexual physiology and behavior in the context of evolutionary biology.

Judson, O. 2003. *Dr. Tatiana's Sex Advice to All Creation*. Owl Books, New York. This is the rarest of creatures: a science book that's at once erudite, informative, and a hoot. Judson's shtick is that she's a sex-advice columnist for animals. And she uses this device to get at some rather sophisticated and subtle issues in the evolutionary biology of sex. Recently, this book has spawned a series of three television episodes featuring Dr. Tatiana, produced by Discovery Channel Canada. These feature elaborate costumed musical numbers, such as "Pocket Rocket" about the evolution of penis shape, that have to be seen to be believed.

Le Vay, S. 1993. *The Sexual Brain*. MIT Press, Cambridge. This is a clear account of the state of brain sex research by a prominent neuroanatomist. The problem is that it's getting somewhat outdated because quite a bit has happened in the field since the time of its writing, and an update is overdue.

SCIENTIFIC REPORTS AND REVIEWS

Allen, L. S., and Gorski, R. A. 1992. Sexual orientation and the size of the anterior commissure in the human brain. *Proceedings of the National Academy of Science of the USA* 89:7199–7202.

Arnow, B. A., Desmond, J. E., Banner, L. L., Glover, G. H., Solomon, A., Polan, M. L., Lue, T. F., and Atlas, S. W. 2002. Brain activation and sexual arousal in healthy, heterosexual males. *Brain* 125:1014–1023.

Bailey, J. M., Dunne, M. P., and Martin, N. G. 2000. Genetic and environmental influences on sexual orientation and its correlates in an Australian twin sample. *Journal of Personality and Social Psychology* 78:524–536.

Bartels, A., and Zeki, S. 2000. The neural basis of romantic love. *NeuroReport* 11:3829–3834.

Chuang, Y. C., Lin, T. K., Lui, C. C., Chen, S. D., and Chang, C. S. 2004. Tooth-brushing epilepsy with ictal orgasms. *Seizure* 13:179–182.

Holstege, G., Georgiadis, J. R., Paans, A. M., Meiners, L. C., van der Graaf, F. H., and Reinders, A. A. 2003. Brain activation during human male ejaculation. *Journal of Neuroscience* 23:9185–9193.

Hu, S., Pattatucci, A. M., Patterson, C., Li, L., Fulker, D. W., Cherny, S. S., Kruglyak, L., and Hamer, D. H. 1995. Linkage between sexual orientation and chromosome Xq28 in males but not in females. *Nature Genetics* 11:248–256.

Karama, S., Lecours, A. R., Leroux, J. M., Bourgouin, P., Beaudoin, G., Joubert, S., and Beauregard, M. 2002. Areas of brain activation in males and females during viewing of erotic film excerpts. *Human Brain Mapping* 16:1–13.

Mustanski, B. S., Dupree, M. G., Nievergelt, C. M., Bocklandt, S., Schork, N. J., and

Hamer, D. H. 2005. A genomewide scan of male sexual orientation. *Human Genetics* 116:272–278.

Pillard, R. C., and Weinrich, J. D. 1986. Evidence of familial nature of male homosexuality. *Archives of General Psychiatry* 43:808–812.

Young, L. J., and Wang, Z. 2004. The neurobiology of pair bonding. *Nature Neuroscience* 7:1048–1054.

7. Sleeping and Dreaming

MATERIAL FOR A GENERAL AUDIENCE

Martin, P. 2003. *Counting Sheep: The Science and Pleasures of Sleep and Dreams.* Flamingo, London. A longish, detailed read but worth the effort. Clearly written, comprehensive, and accurate.

Rock, A. 2004. *The Mind at Night.* Basic Books, New York. This book is more focused on dreams than sleep as a whole. It relies heavily on interviews with a group of prominent sleep researchers. Rock is at her best when she is telling some of the personal stories behind the science.

SCIENTIFIC REPORTS AND REVIEWS

Frank, M. G., Issa, N. P., and Stryker, M. P. 2001. Sleep enhances plasticity in the developing visual cortex. *Neuron* 30:275–287.

King, D. P., and Takahashi, J. S. 2000. Molecular genetics of circadian rhythms in mammals. *Annual Review of Neuroscience* 23:713–742.

Louie, K., and Wilson, M. A. 2001. Temporally structured replay of awake hippocampal ensemble activity during rapid eye movement sleep. *Neuron* 29:145–156.

Nikaido, S. S., and Johnson, C. H. 2000. Daily and circadian variation in survival from ultraviolet radiation in *Chlamydomonas reinhardtii*. *Photochemistry and Photobiology* 71:758–765.

Pace-Schott, E. F., and Hobson, J. A. 2002. The neurobiology of sleep: genetics, cellular physiology, and subcortical networks. *Nature Reviews Neuroscience* 3:591–605.

Ribiero, S., Gervasoni, D., Soares, E. S., Zhou, Y., Lin, S.-C., Pantoja, J., Levine, M., and Nicolelis, M. A. L. 2004. Long-lasting novelty-induced neuronal reverberation across slow-wave sleep in multiple forebrain areas. *PLoS Biology* 2:126–137. *PLoS Biology* is an open-access journal whose contents are available free to all at www.plos.org.

Siegel, J. M. 2005. Clues to the function of mammalian sleep. *Science* 437: 1264–1271. This review is highly critical of the REM sleep/memory consolidation hypothesis. Reading it together with Stickgold 2005 will give you both sides of the argument.

Stickgold, R. 2005. Sleep-dependent memory consolidation. *Science* 437:1272–1278.

Wagner, U., Gais, S., Haider, H., Verleger, R., and Born, J. 2004. Sleep inspires insight. *Nature* 427:304–305.

8. The Religious Impulse

MATERIAL FOR A GENERAL AUDIENCE

Boyer, P. 2001. *Religion Explained.* Basic Books, New York. A cognitive anthropologist examines the question "Why do we have religion at all?" from a cross-cultural and evolutionary perspective.

Brockman J., ed. 2006. *What We Believe but Cannot Prove: Today's Leading Thinkers on Science in the Age of Certainty.* New York, Harper Perennial.

Gazzaniga, M. S. 1998. *The Mind's Past.* University of California Press, Berkeley. A fun, brief book laying out the case for a specific module in the left brain for interpreting disparate data and constructing narratives. Written in a conversational style, Gazziniga's book lets his quirkiness and wit shine through. Don't believe the short shrift he gives to experience-dependent plasticity though. In arguing (appropriately) against the blank-slate behaviorist tradition, he got a bit carried away.

9. The Unintelligent Design of the Brain

MATERIAL FOR A GENERAL AUDIENCE

Brockman, J., ed. 2006. *Intelligent Thought: Science versus the Intelligent Design Movement.* New York, Vintage. This is a collection of essays by prominent scientists refuting the intelligent design model. The essay by Jerry Coyne is the best and most succinct argument based on the fossil record that I know.

Pennock, R. T., ed. 2001. *Intelligent Design, Creationism, and Its Critics.* MIT Press, Cambridge. This large volume is a good way to get started if you are highly motivated to hear the arguments on both sides of this contentious issue.

Acknowledgments

I HAVE BEEN BLESSED to work in a stimulating and dynamic environment. This intellectual milieu has been central to shaping my thoughts as expressed in this book. First and foremost, I would like to thank my wife, Professor Elizabeth Tolbert, who is, quite simply, the smartest and most interesting person on the planet. A true scholar and a fearless thinker, she has spent years pushing and prodding me to take a few baby steps outside of the scientific mainstream. Most of the ideas in this book have been stimulated by our ongoing discussions (characterized by our friends with a Zen-like bent as "the sound of two rocks crashing together.")

A group of brilliant and friendly colleagues at The Johns Hopkins University School of Medicine continue to make my work life a joy. In particular, I owe a debt of gratitude to the Neuroscience Lunch Crew: David Ginty, Shan Sockanathan, Alex Kolodkin, Rick Huganir, Dwight Bergles, Paul Worley, and Lunch Crew Emeritus Fabio Rupp, who have provided both intellectual and social support over many years. The people who have worked in my lab continue to humble me with their insights, hard work, and friendship. Thanks to Kalyani Narsimhan, Kanji Takahashi, Carlos Aizenman, Christian Hansel,

Angèle Parent, Dorit Gurfel, Shanida Morris Nataraja, Jung Hoon Shin, Ying Shen, Andrei Sdrulla, Yu Shin Kim, Wei Zhang, Roland Bock, Hiroshi Nishiyama, Sang Jeong Kim, Sangmok Kim, and Joo Min Park.

Many thanks to Sol Snyder, Department Director Extraordinaire, who has been supportive in all things. This book was written during a lovely sabbatical year. My academic home away from home was Wolfson College, University of Cambridge. In particular, I would like to thank Ian Cross and Jane Woods for going above and beyond the call of duty in welcoming me and my family to Cambridge. Thanks also to the scientists of the Department of Physiology at University College, London, and Paola Pedarzani in particular, for encouraging and tolerating my frequent visits on seminar days.

A number of good people provided advice and critiques of various parts of the manuscript. I am indebted to Elaine Levin (Mom!), Keith Goldfarb, Sascha du Lac, Eric Enderton, Steven Hsiao, Nely Keinanen, Herb Linden (Dad!), Sue Reed, Julia Kim Smith, and The Prince of Dark Moods himself, Adam Sapirstein. Many scientists took time from their busy schedules to prepare figures or track down obscure information. My thanks to Niko Troje, Kristen Harris, Anthony Holtmaat, Yao-Chung Chuang, Ullrich Wagner, Frank Schieber, and that amazing web detective, Roland Bock.

Many professionals in the publishing world have lent their talents to this effort. Joan M. K. Tycko took my horrid sketches and half-baked graphic ideas and transformed them into superb illustrations. Michael Fisher, Editor-in-Chief at Harvard University Press, has been insightful and supportive throughout the publishing process. Nancy Clemente labored hard to clean up my clunky prose.

Thanks to Cal Fussman for generously allowing me to use a quotation from his interview with Bruce Springsteen, which originally appeared in *Esquire* magazine on August 1, 2005. And thanks to the University of Wisconsin Press for allowing me to reprint a quotation by Donald O. Hebb.

Finally, I couldn't go on without the love and inspiration of Jacob Linden and Natalie Linden.

Index

child abuse, accusations of, 126
childhood, long, 81
child rearing, 149–150, 152; paternal involvement in, 146–148
children, suggestibility of, 126
chimpanzee, 60–61
chloride ions, role in electrical signaling, 45
Cho, Margaret, 146
cholinergic neurons, 206–207, 214
chronostasis, 97
Chuang, Yao-Chung, 171
circadian clock, 203–206
classical conditioning, 110–111, 249–250
climbing fibers, 251
clonezepam (Klonopin), 192
Cnidaria, 29
cocaine, 164
cognitive ability, variation in, 24–26
cognitive style, gender differences in, 157–162
cold/menthol receptors, 248
"cold," perception of, 248
complexity, irreducible, and intelligent design, 238–242
conditioned place aversion, 102–103
confabulation, 226
congenital adrenal hyperplasia, 158–159, 181
continuity, in sensory processing, 94–97
coordination of movement, 9
corpus callosum, 155, 159–160, 226–227
cortex, 7, 18–21, 61, 64–65, 80, 116; anterior cingulate, 101–104, 163, 165, 215; association, 18–19, 24; cerebral, 155, 164; frontal, 19, 21, 166, 170; left, 227–229; motor, 18; occipital, 166; parietal, 170; prefrontal, 119, 215; primary auditory, 84; primary somatosensory, 84–86, 101; primary visual, 83–84, 214; right, 227–229; somatosensory, 10; temporal, 166, 170
cortisol (hormone), 186
Coyne, Jerry, 236
cranial nerves, 7
cross-dressing, 153
cultural war, between science and religion, 232–234
culture, and gender identity, 154

curare, 45
cyanobacteria, 205
cytokines, 66

Dalai Lama, 234
Darling, Sir Frank, 148
dead phone illusion, 97
deaf humans, and brain wiring, 71
death: during childbirth, 151; from sleep deprivation, 186–187
declarative memory, 109, 114–117, 121, 249
delay, in awareness of sensory information, 96
delayed matching to sample task, 117–119
Delbrück, Max, 4
Dembski, William, 238, 245
dendrites, 29, 65, 73
dendritic spines, 29
DES (diethylstilbestrol), 158–159
Diamond, Marion, 75–77
DNA, 56–58, 176–178. *See also* genes, human; genetic factors; human genome
dopamine (neurotransmitter), 47, 119
dorsal raphe (brain region), 207, 215–216
dreaming, 213–219. *See also* dreams
dream interpretation, 207–208, 220
dream journals, 209–210, 212
dream research, 188
dreams: function of, 216–219; meaningful/symbolic, 207–208, 219–220; narrative, 211–213, 229–232 (*see also* religious thought); sexual, 212; sleep-onset, 210
drugs. *See names of drugs*
drug use, maternal, 66–67

ECT (electroconvulsive shock treatment), 119–121
ectoderm, 59
edge enhancement, in visual system, 93
EEG recordings of sleep, 189
Einstein, Albert, 44; brain of, 25–26
Eisenberger, Naomi, 103
ejaculation, male, 169–170
Elavil (antidepressant), 201

electrical activity, and brain wiring, 74–75
electrical signaling in brain, 32–44
embryonic disk, 59
emotion: and amygdala, 16–17; and dreams, 212, 215, 219; and limbic system, 16; and memory, 107–109, 122–123; and pain, 100–104; and perception, 97–104
empathy, 103–104
energy conservation, and sleep, 187–188
engineering of brain. *See* brain design
environmental deprivation, 77
environmental enrichment, 77, 79–80
environmental factors, 52, 80–81; in brain development, 52, 66, 74–80; in brain wiring, 71–72. *See also* nature-nurture debate
enzymes, 56; acetylcholinesterase, 56; protein kinase C, 251
epigenetic factors: in brain development, 52 (*see also* nature-nurture debate); and sexual orientation, 178
epilepsy, 17, 64, 112–114, 226–227
EPSP (excitatory postsynaptic potential), 39–40, 44
"escape from light" hypothesis, 205–206
estrogen (hormone), 58, 167
evolution: of brain design, 6, 21–22, 26–27; of brain size, 24, 61; of circadian clock, 205–206; of dreams/dreaming, 217; and intelligent design, 235–246; and memory storage, 143–144; of neurons, 29; and REM sleep, 192–194; and sensory processing, 92–97; and sexual behavior, 152
evolutionary biology, 233–238
excitation (in neural signaling), 44
experience: and gender identity, 154; molecular/cellular underpinnings of, 248–254

face-blindness (prosopagnosia), 88
false memories, 125–126
fear: amygdala and, 16–17; in dreams, 211–213, 215, 217, 219
females, and sexual behavior, 167–168
fetal vision, 70
fight-or-flight responses, 17, 100

Firestein, Harvey, 174
flagellum, 239
fMRI (functional magnetic-resonance imaging), 10, 104
food preferences, and nature-nurture debate, 54
forgetting curve, 119
Foulkes, David, 230
founder effect, 79
Freud, Sigmund, 208
frog brain, 21–22
frogs, 67–70
frontal cortex, 19, 21, 166, 170
fruit fly, 60
fugu (pufferfish), 41
fundamentalism, religious, 233–238
fungi, 205

GABA (gamma-aminobutyric acid), 45, 47, 93, 192
Galton, Francis, 90
Gardner, Randy, 186
gays, 173–182
Gazzaniga, Michael, 227, 229
gender dysphoria, 153
gender identity: development of, 153–160; diversity of, 173–174. *See also* sexual orientation
gene expression, 56–59
general intelligence: genetic factors, 175, 177; lack of gender difference, 157; tests of, 53–54
genes, human, 51–52, 56; ASPM gene, 60–61, 243; homeotic, 61–64
genetic factors: in brain development, 52, 59–61, 80–81 (*see also* nature-nurture debate); in gender differences, 157–162; in general intelligence, 175, 177; in sexual orientation, 175–182
genetic specification, in brain wiring, 71–72
genitals, and tactile sensitivity, 84–86
genome, and evolution, 108, 242–243. *See also* human genome
glial cells, 28; radial glia, 64
glutamate (neurotransmitter molecule), 35–37, 44–47, 65, 103, 251

glutamate receptor proteins, 37–39, 56
glycine (neurotransmitter), 45
gorilla, 61
Gorski, Roger, 179
Granholm, Jackson, 6
granule cells, 251
gray matter, 60
Green, Richard, 180
growth hormone releasing hormone, 16
guidance molecules, in brain development, 67–70

hallucinations, 186, 215
Hamer, Dean, 177
Hartmann, Ernest, 217
hearing, 13–14, 71, 84
"heat," perception of, 247–248
Heisenberg, Werner, 44
heroin, 164
higher cognitive processes, brain-critical periods, 77–80
Hines, Melissa, 159
hippocampal system, 112
hippocampus, 16–18, 248–249; and tests of memory storage, 132–143
Hobson, J. Allen, 219
Hogg, Andrew, 186
Holstege, Gert, 170
homeostasis, 15
homeotic genes, 61–64
homosexuality, 148, 173–182
hormones: circulating, 66; cortisol, 186; estrogen, 58, 167; growth hormone releasing hormone, 16; master, 16; oxytocin, 173; progesterone, 167; secreted by hypothalamus, 16; testosterone, 158, 168–169; thyroid, 58; vasopressin, 16
"housekeeping genes," 56
Howe, Elias, 196
Hubbard, Edward, 91
human genome, 56
Human Genome Project, 51–52
hunger, 15
hydraulic analogy, for electrical signaling, 42

hypnotic suggestion, and pain modulation, 102
hypothalamus, 15–16, 166, 173; lateral nucleus, 15; medial preoptic region, 168–169; SCN (suprachiasmatic nucleus), 203, 206–207; ventromedial nucleus, 15, 167

immune system, mother's, 66
Imperato-McGinly, Juliane, 159
implicit memory. See nondeclarative memory
INAH3 (interstitial nucleus of the anterior hypothalamus number 3), 155, 168, 178–179
individuality, human, and brain development, 81
inferior olive, 250
inhibition (in neural signaling), 44–45
inhibitory synaptic drive, and REM sleep, 191–192
insight, sleep-inspired, 196–197
insula (brain region), 101, 163, 165
intelligence testing, 25, 53–54
intelligent design, and evolution, 235–246
interview studies, of sexual orientation, 180
intrinsic plasticity, 143
ion channel, 38–40, 56
IPSP (inhibitory postsynaptic potential), 45

Jacob, François, 6
James, William, 53
Jäncke, Lutz, 89
jellyfish, 29
jet lag, 204–205
John Paul II, Pope, 235–237
Johnson, Carl, 206
Johnson, Philip E., 237–238, 245
Jouvet, Michel, 192

Karama, Sherif, 166
Kekulé, Friedrich, 196
Kleitman, Nathaniel, 189
Klinefelter's syndrome, 176
kludge, brain as, 6, 22, 48–49, 240–242, 245–246

language: acquisition of, 77–78; and mirror neurons, 105
lateral inhibition, 93
lateral nucleus (of hypothalamus), 15
learning: and memory, 133, 138–143, 248–249; and sleep deprivation, 196–197; tests of, 250–254
left cortex, and split-brain operation, 227–229
Lenin, V. I., brain of, 25
lesbians, 173–182
LeVay, Simon, 178–179
limbic system, 16
Linné, Carl von, 205
localization, of brain functions, 21–24
local signals, in neuronal diversity, 64–65
locus coeruleus, 207, 215–216
Lomo, Terje, 133–134
Louie, Kendall, 197–200
love, neurobiological basis for, 162–166
LTD (long-term synaptic depression), 134–143, 248–249, 251
LTP (long-term synaptic potentiation), 134–143, 248–249

macaque monkey, 61
mahu, 154
males, and sexual behavior, 168–169
malnutrition, maternal, 80
master hormones, 16
masturbation, 148
mathematical reasoning, and gender, 157
McCartney, Paul, 196
M cells (of the visual system), 86–89
medial preoptic region (of the hypothalamus), 168–169
melanopsin-positive ganglion cells, 203–205
melatonin, 206
membrane potential, 39
memory: and brain development, 81; declarative, 109, 114–117, 121, 249; and emotion, 107–109, 122–123; for facts and events, 17–18; false, 125–126; and hippocampus, 17; and learning, 133, 138–143, 248–249; and limbic system, 16; long-term, 119–121;

nondeclarative, 112, 114–116, 121, 250; short-term, 119–121; taxonomy of, 109; working, 117–119
memory consolidation, 119–121, 123, 197, 208, 217–219
memory duration, 116–121
memory integration, 218–219
memory localization, 112–116
memory retrieval, 121–127
memory storage, 112–116, 126–132, 195–202; tests of, 132–143
menopause, 147
mental function, brain's creation of, 48–49
mental retardation, 64
mice, 70–71, 132, 138, 141
microcephaly, 60–61, 243
midbrain, 13–15, 61, 67
mirror neurons, 105
mirror reading, 109–110
misattribution, and memory retrieval, 123–124
mitotic spindle, 60, 243
molecular genetics, 55–59
monkeys: and female sexual circuit, 167–168; and memory tasks, 117–119
monoamine oxidase inhibitors, 201
monogamy, 146–147
mood regulation, and dreaming, 216–218
mossy fibers (of the cerebellum), 250–251
mother-infant bond, 173
motor coordination learning, 110
motor cortex, 18
motor function, and sensation, 104–105
mutation, random, and "specified complexity," 239–240
myelin secretion, 72–73

Nadel, Lynn, 141
narrative creation, propensity for, and religious thought, 225–232. See also under dreams
Native American culture, 154
nature-nurture debate, 53–59, 80–81; and gender differences, 157–162; and sexual orientation, 173–182
Neanderthal man, 24

neural Darwinism, 74–75

neural plasticity, 75–80

neural plate, 59

neural tube, 59, 61, 64

neurology, 97–98

neuronal activity, and brain wiring, 70–71

neuronal cell culture, 35

neuronal diversity, 64–65

neuronal migration, 64

neuronal plasticity, 80–81

neurons, 28–29; and accidental design, 242; cholinergic, 206–207, 214; mirror, 105; number of, 73–74, 81; requiring multiple simultaneous synapses, 45; role in electrical signaling, 34–44; temporal limits of spike firing, 43, 48

neurosurgery, 112–114

neurotoxins, 40–41

neurotransmitter receptors, 33, 56; glutamate receptor proteins, 37–39, 56; NMDA-type glutamate receptor, 135–141, 248

neurotransmitters, 32, 37, 47; acetylcholine, 45, 47, 56, 206, 215–216; dopamine, 47, 119; fast, 45–46; glutamate, 35–37, 44–47, 65, 103, 251; glycine, 45; noradrenaline, 46–47, 215–216; slow-acting, 46

nicotine use, maternal, 66–67

Nikaido, Selene, 206

NMDA-type glutamate receptor, 135–141, 248

nondeclarative memory, 112, 114–116, 121, 250

non-REM sleep, 189–194, 200, 207, 210–211

Noonan, Katherine, 149

noradrenaline (neurotransmitter), 46–47, 215–216

occipital cortex, 166

O'Keefe, John, 141

olfaction, and sexual behavior, 168

optical illusions, 93, 96–97

oral-genital stimulation, 148

orangutan, 61

orgasm, 169–173

outer membrane (plasma membrane) (neuron), 29

ovulation, 147; concealed, 149, 151

oxytocin (hormone), 173

pain, and emotion, 100–104

pain asymbolia, 101–102, 173

pair bonds, 146–147, 149–151, 173

parallel fibers, 251, 253

paralysis, limp muscle, in REM sleep, 191, 216

paranoia, caused by sleep deprivation, 186

parietal cortex, 170

parietal lobe, 87

paternal involvement in child rearing, 146–150

paternity, of offspring, 147–148, 151–152

Pavlov's dog, 249–250

P cells (of visual system), 86–89

Penfield, Wilder, 112–114

perception/emotion distinction, 97–104

personality change, from damage to cortex, 19–21

PET scanning (positron emission tomography), 170, 214–216

phenelzine (Nardil), 201

phrenology, 22

pineal gland, 206

Pittendrigh, Colin, 205

pituitary gland, 173

place cells, 141, 197–200

plagiarism, 124

plasticity: neural, 75–80; neuronal, 80–81; synaptic, 49, 143

playback of memories, 197–200, 212–213

pleasure, sensations of, 164. *See also* orgasm

Pliny the Elder, 205

police line-ups, 125

polyandry, 147

polygenic traits, 177

polygyny, 147

Polynesian culture, 154

pontine tegmentum, 214

postmortem studies, 178–179

post-orgasmic afterglow, 173

potassium, in cerebrospinal fluid, 34–35

potassium ions, role in electrical signaling, 38–40

PP1 (protein phosphatase 1), 137

signal relay, thalamus and, 15

Skinner, B. F., 53

sleep: non-REM, 189–194, 200, 207, 210–211; physiological functions of, 187–188; REM (rapid eye movement), 189–194, 197–202, 207, 209, 211–213

sleep cycle, 189–191, 206–207, 211; changes in, 192–194; function of, 194–196

sleep deprivation, 184–187, 195; and learning, 196–197; non-REM sleep, 200; REM sleep, 197

sleep research, 188–191

sleepwalking, 192

snails, 29

social drives, hypothalamus and, 16

social interaction, and sensory systems, 103–104

sodium, in cerebrospinal fluid, 33–35

sodium ions, role in electrical signaling, 38–40

somatosensory cortex, 10

source misattribution, and memory retrieval, 123–124

spatial learning, as test of memory storage, 133, 138–143

spatial skills, and gender, 157

Sperry, Roger, 67, 227

spike (electrical signal), 32, 39–40, 127

spike firing: and memory storage, 127–129; pattern of, 43, 48

Spitzer, Robert, 181–182

splenium (subregion of corpus callosum), 159–160

split-brain patients, 226–229

SRE (gene promoter), 58

SRF (a transcription factor), 58

SSRIs (serotonin-specific reuptake inhibitors), 201

Stickgold, Robert, 195, 210

stopped-clock illusion, 96–97

stress: maternal, 80; from REM sleep deprivation, 200–201

strychnine, 45

subconscious, and dreams, 208

Südhof, Thomas, 70

suggestibility, and memory retrieval, 123–126

Summers, Larry, 160–162

Svoboda, Karel, 132

symmetry of brain, 7

synapses, 30–33; number of, 32; probabilistic function of, 44, 48

synaptic cleft, 32

synaptic competition, 74

synaptic connections, and synesthesia, 91

synaptic depression, 130

synaptic function, experience-dependent modification of, 129–131

synaptic nametags, 67–69, 81

synaptic plasticity, 49, 143

synaptic potentiation, 130

synaptic strength, and memory storage, 129–131

synaptic structure, changes in, and memory storage, 131–132

synaptic vesicles, 32, 43

synesthesia, 89–92

tactile form perception, 86

temporal cortex, 166, 170

testosterone (hormone): and male sexual behavior, 168–169; prenatal, 158

tetrodotoxin, 41

thalamus, 15, 101, 206

theory of mind, 24, 105

thirst, 15

thyroid hormone, 58

tickling, 10–12

timing: of REM sleep, 197; of sexual intercourse, 146–147; of sleep-wake cycle, 203–206

tit-for-tat experiment, 12–13

torture by sleep deprivation, 184–187

touch, 84–86, 102

transcription factors, 58, 61

tricyclic antidepressants, 201

twins, and brain development, 66

twin studies, 53–55; identical twins, 53–55, 60; nonidentical twins, 55, 60; of sexual orientation, 175–176

two-spirit (Native American cultural practice), 154

Uncertainty Principle (Heisenberg), 44

Valium, 201
vanilloid receptors, 248
vasopressin (hormone), 16
ventral tegmental area, 164, 166, 170
ventricles, 59
ventromedial nucleus (of hypothalamus), 15,
   167
verbal fluency, and gender, 157
Versed (a sedative drug), 201
vision, midbrain and, 13–14
visual object agnosia, 88
visual system, 80; and circadian clock, 203–205;
   and coherent image, 225; development of, 67–
   70, 75; and edge enhancement, 93; and map
   of visual world, 83–84; organization of, 86–89
vital functions, controlled by brainstem, 9

wakefulness and sleepiness, brainstem and, 9. *See*
   also sleep cycle

Warrington, Elizabeth, 111
weight loss, 15
Weinrich, James, 180
Weiskrantz, Larry, 111
"what" pathway, in visual system, 87–89
"where" pathway, in visual system, 87–89
White House Conference on Early Brain Development (1997), 78–79
white matter, 60, 73
Wilson, Matt, 197–200
Winson, Jonathan, 217
Wolpert, Daniel, 10, 12
working memory, 117–119
worms, 29

Xanax (antiaxiety drug), 201
X chromosome, and male sexual orientation,
   176–178
XXY genotype, male, 176

Zeki, Semir, 162–164